The Art of Lying Down

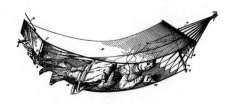

Also by Bernd Brunner

Inventing the Christmas Tree
Moon: A Brief History
Bears: A Brief History
The Ocean at Home: An Illustrated
History of the Aquarium

THE Art OF Lying Down

A GUIDE to HORIZONTAL LIVING

by BERND BRUNNER   translated by LORI LANTZ

MELVILLE HOUSE
BROOKLYN • LONDON

The Art of Lying Down

Originally published in the German language as
*Die Kunst des Liegens: Handbuch der horizontalen
Lebensform* by Bernd Brunner

Copyright © 2012 by Verlag Kiepenheuer &
Witsch GmbH & Co. KG, Köln / Germany

Translation copyright © 2013 by Lori Lantz

First Melville House printing: November 2013

Melville House Publishing   8 Blackstock Mews
145 Plymouth Street   and   Islington
Brooklyn, NY 11201   London N4 2BT

mhpbooks.com   facebook.com/mhpbooks   @melvillehouse

ISBN: 978-1-61219-309-0

Manufactured in the United States of America
1 3 5 7 9 10 8 6 4 2

Design by Christopher King

A catalog record is available for this book
from the Library of Congress.

A thing that can't be done in bed isn't worth doing at all.

— Groucho Marx

Contents

3	Are You Lying Down?
7	The Grammar of Horizontal Orientation
10	Drawn to the Center of the Earth
12	Chesterton and the Secret of Michelangelo
14	Shaking Up the Act of Lying Down
17	Common and Uncommon Ways to Lie Down
21	Lying Down in the Great Outdoors
25	Sun Worshippers
27	The Proper Way to Lie Down
33	Position as the Key to Personality
35	So Easy a Child Can Do It
38	Lying Down Together
42	Lying Down, Sleeping, Waking Up
52	Awake, Napping, Asleep
56	Eating and Lying Down: Better Together?
60	Horizontal—but Hard at Work
70	The History of the Mattress

77   The Archaeology of Lying Down

92   The Oriental Roots of the Art of
      Lying Down

96   Field Studies of Bedrooms and
      Reclining Habits

101   The Typical Bed

105   Lying Down on the Road

112   Strange Bedfellows

120   Mechanized Reclining

126   Horizontal Healing

130   Floating, Rocking, Swinging

135   The Puzzle of the Recliner

138   The Best Place for the Bed

143   Lying Down as the Stuff of
      Dreams—and Nightmares

149   The Museum of Reclining

152   Are You Still Lying Down?

157   For Further Reading

161   Illustration and Photo Credits

163   Index of Names

167   Acknowledgments

The Art of Lying Down

Are You Lying Down?

If you're lying down right now, there's no need to defend yourself. We all do it regularly, and often we enjoy it. We lie down to relax, assuming the posture that offers the body the least resistance and demands the least energy. And we perform all sorts of activities this way: we sleep and dream, make love, contemplate, give ourselves over to wistful moods, daydream, and suffer. But there's one thing we rarely do in this state: move around. When we stretch out horizontally, we come the closest we can to remaining still.

In a society attuned to measurable performance, where quickly making and acting on decisions are what matters and people prove themselves by sitting for long hours at their desks and in front of their computers, reclining often goes unappreciated. Even worse, it is seen as proof of indolence or a sign of powerlessness in a fast-changing world. You can't keep up when you're lying down. Those who do anyway are considered weak or criticized for not putting their time to better use. Yet lying down can feel like taking a walk in a thick fog: we often emerge with clearer thoughts than before. As a calculated move to escape the ever-present pressure to be fast and efficient, conscious reclining costs nothing and is yet extremely valuable.

Lying down is the horizontal counterpart of the dreamy rambling of a melancholy flâneur, who walks about without pursuing any goal. Someone in repose may wander through town and countryside, too, but generally only in his or her imagination. These fanciful strolls demand a higher level of creativity since no real faces and places show up to stimulate the resting wanderer's thoughts.

When we lie on our back and direct our gaze up toward the ceiling or sky, we lose our physical grasp of things and our thoughts soar. Our mental makeup and even the structure of our perception can change with this shift of position. Responses that seemed perfectly natural a few minutes earlier, when we were standing upright, become inexplicable. Questions that were so important appear in a different light when we view them horizontally. Voices and

BERND BRUNNER

even the ringing of a telephone may no longer reach us with the same urgency as when we are standing. In no other position can certainties suddenly seem less certain. When we lie down, perhaps because we feel overwhelmed, a burden falls from our shoulders. But perception patterns differ, possibly even from person to person. Lin Yutang, the Chinese writer, once claimed that "our senses are the keenest in that moment" when we are lying down, and he goes on to say that "all good music should be listened to in the lying position."

Thinking about what it means to recline involves not only questions of physiology, psychology, and creativity but also the economy of time and the pace of our lives, which the American psychologist Robert Levine once tellingly described as "a tangled arrangement of cadences, of perpetually changing rhythms and sequences, stresses and calms, cycles and spikes." If and when lying down is acceptable all depends on the attitudes towards time. We operate in time, and it governs our behavioral cycles like a silent language. In an age and culture like ours, which has internalized a compulsion for constant movement and bred an internal agitation that rules every aspect of our lives, there is little we can do but turn the screws of time and adapt to its demanding rhythm. In places with different rhythms from our own, where activities emerge from what's happening at the moment rather than from what has been planned, we can get a sense of what it means to live within time that follows different laws. In other societies, a moment when nothing seems to be happening may not be seen as a "waste of time," but rather as something pleasant and essential to life. Another reason to give the horizontal world a closer look.

The Grammar of
Horizontal Orientation

Our bodies are designed for performance well beyond the limited movements we demand of them. We spend far more time sitting and move much less than our ancestors did just a few generations ago. Thanks to our genetic makeup and physical disposition, we are born to vary our movements, from walking to lying down, standing, sitting, and more. Reclining horizontally is just one of these many possibilities, but the urge to give in to the pull of gravity is a strong one. Gravity pulls us toward the earth, and we are constantly negotiating its demands. Although we aren't consciously aware of this effort—we are so accustomed to it that we do it automatically—a great deal of our energy is dedicated to our struggle against this elemental physical force.

Lying down and walking or running are counterparts to each other, since one determines the other. Only someone who has walked, hiked, or run to the point of exhaustion can fully appreciate the infinite sense of rest that lying down can bring. Others never fully experience this sensation. But reclining can also present a different purpose: it offers a retreat when the body becomes too heavy to bear. Lying down is like powering down to zero.

We orient ourselves in two directions from the earth's surface: vertical and horizontal. When we walk, our connection to the earth is experienced through the soles of our feet, but when we lie down, this contact spreads to the rest of our body. When we are upright, the horizon, the distant line separating sky and earth, promises something beyond what we know but also represents a limit reaching that beyond. It is a fleeting construct that retreats into the distance whenever we try to approach it, a goal that can never be reached.

Lying down is possible to execute in many locations. While a cot or a bed is not essential, reclining does require a stable foundation. If we can't lie down comfortably and are worried about our safety—perhaps because we're in danger of falling off the surface, or rolling off—we won't be able to relax. For this reason, the preparations we make for being horizontal shape how we experience this (in)activity. In the end, a person's chosen method of reclining is a response to a particular resting place. The more comfortable we make ourselves by providing the best support for our weight, the greater the renewal we will feel.

Defining what it means to lie down is not as easy as it might seem. One might define it as when most

BERND BRUNNER

of a body is in a horizontal position, or clearly tending in that direction, and is shifting the body's weight to the underlying surface. It is possible to lie on the back ("supine"), the sides, or stomach ("prone"). We can lie next to, on top of, or—at least for a short time—under another person. Lying down can also take the form of leaning the torso back horizontally and raising the legs higher than the torso. Le Corbusier's chaise longue, in which the upper body rests at a forty-five-degree angle, is designed for this sort of reclining.

Unlike sitting in a chair, which requires some physical control, lying down requires no effort at all. It is therefore perhaps the most archaic of all positions and reminds us of earlier states of existence. But standing up, especially if there is no pressing need to do so, means overcoming gravitational resistance. Lying down does entail a certain risk, though, because it's easy to drift off into unconsciousness.

A great deal can happen when we are lying down. This position spans the human condition, from complete passivity to the most passionate of activities. Furthermore, human life begins and ends horizontally. According to Edmond and Jules de Goncourt, the "three great acts of life" that a writer must master are "birth, coitus, and death," all of which usually involve lying down.

Drawn to the Center of the Earth

We tend to consider someone lying down to be passive, paralyzed, or at the power of others. Of course, such an impression often bears little relation to the motivations of the reclining person. Perhaps he wants to let go, rest, and relax; perhaps she wants to gather her energies for a next move. For someone lying in wait, lying down can form part of a clever strategy. It can also be an act of rebellion, as when large numbers of people come together and lie down to block passersby or traffic. These examples nicely contradict the statement Elias Canetti once made: "A man who lies down gives up all relationships with his fellows and withdraws into himself."

Reclining is also the preferred posture of the lazy. As the German writer Hans W. Fischer once wrote:

> Yet utter laziness seeks after nothing: no joy, not even complacence. It does not occur to laziness to make the slightest preparation to enjoy itself. Instead, it simply lets itself collapse and—as long as a wall does not happen to be in the way—follows the laws of mechanics to end up in an approximately horizontal position.

Its preferred spot is the sofa, because it is so convenient; a dim remnant of consciousness warns it away from the naked floor, which would hurt to fall on, and keeps it from the bed, which harbors associations with the complicated act of getting undressed. But laziness does not seek out the sofa to sleep or enjoy a luxurious stretch; no, it simply needs a landing place for the weight of a body that feels drawn toward the center of the earth.

Being tired often seems like the only acceptable reason to lie down. Why is this pleasurable position so frowned upon? All too frequently we have internalized the sense that we have to be moving at every moment, that anything less reveals a lack of discipline, strength, and ambition. In a world that demands we stay on the go and make the most of our time, where flicking off the office lights late in the evening is a source of pride, time spent lying down appears to be time wasted. In our culture, lingering in the horizontal is acceptable only for the shortest possible period required to power the next bout of activity.

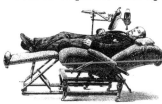

American patent for recumbent total care

Chesterton and the Secret
of Michelangelo

Is there such a thing as a philosophy of lying down? Many have expressed their disapproval of staying in bed, dismissing this behavior as nothing but senseless laziness, while others have simply practiced the rite of leisure without feeling the need to explain themselves. But has anyone actually thought deeply about the act of lying around and had something positive to say about it? One man in particular did just that: the prolific English social critic Gilbert Keith Chesterton (1874–1936). His essay "On Lying in Bed" begins with a thought experiment: he imagines how nice it would be to have colored pencils long enough to use to draw on the ceiling while lying in bed. After all, only the ceiling offers a surface large enough for artwork; all the walls are covered already with wallpaper. His thoughts then turn to Rome: "I am sure that it was only because Michael Angelo was engaged in the ancient and honorable occupation of lying in bed that he ever realized how the roof of the Sistine Chapel might be made into an awful imitation of a divine drama that could only be acted in the heavens."

He opposes the general disdain for lying down, a

disdain he considers "unhealthy" and "hypocritical," and encourages the freedom and flexibility of each individual to decide for himself when to get out of bed as he sees fit, or to enjoy his lunch "sometimes in the garden, sometimes in bed, sometimes on the roof, sometimes in the top of a tree." Although Chesterton recommends that these bouts of leisure (which, in his description, do not involve sleeping) should be "very occasional," he insists that it is unnecessary to justify such behavior, except in cases of serious illness. "If a healthy man lies in bed," he explains, "let him do it without a rag of excuse; then he will get up a healthy man." However, he continues, "If he does it for some secondary hygienic reason, if he has some scientific explanation, he may get up a hypochondriac."

Shaking Up the Act of Lying Down

Reclining can be a more active or passive activity depending on the mood of those doing it, while the way we work influences how we spend the rest of our time. Those who spend most of their working hours seated—for example, in front of the computer screen—are likely to seek out sports and movement in their free time. Only then can they fully enjoy the relaxation that, later, lying down provides. On the other hand, people who work with their muscles, perhaps even to exhaustion, generally want to spend their leisure time simply relaxing. For them, lying down has a different character.

How do the significant changes taking place in the world of work today affect the relationship of work and free time—in terms of both the various stages of lounging and lying around we practice and the time we spend sleeping? Does a greater degree of flexibility allow for more idleness, more playful ease—more active laziness, if you like—in a kind of lying down that adds up to more than regeneration alone?

In a time when progress occasionally misses its mark by a wide margin, lying down is a preliminary exercise for thinking things through. It also has the pleasing quality of being removed from the compulsive forward-and-backward logic of progress and decline.

Have we forgotten how to lie down, just as we have forgotten how to cook when we eat nothing but takeout meals for too long? Perhaps. Just as eating is more than a way to fuel further work and physical processes, lying down has dimensions beyond those that prepare us to sit at our desks again. For although it's true that reclining rarely produces any directly visible or economically exploitable outcomes, resting is not its sole purpose either.

Evidence exists that our society's attitudes towards lying down are currently undergoing a transformation. In recent years French commentators have noted the rise of the *génération vautrée*, the lolling generation that refuses to sit upright. Instead of sitting down, its members collapse on the sofa or bed and get comfortable in this position without wasting a thought on what those around them may think. An enviable attitude. It's possible to read this laxness as a revolt against sitting still and standing up straight, or it can be seen as a silent protest against parents who may set down authoritarian rules. But physicians are likely unfazed by this development; they are aware that, in physiological terms, a normal position in a chair is actually unsuitable for most people.

Reclining at a 127-degree angle, however, seems to do a good job of eliminating the tension that builds up in the spine when we sit.

Adherents of the art of slow living, or *l'arte del vivere con lentezza*, offer further signs that these attitudes are changing. Like the slow food initiative, this movement has its roots in Italy. It is a movement toward deceleration and reflection, the point being not to achieve a goal in the shortest time and with the least effort, but rather to enjoy the process itself, in keeping with the saying "The journey is its own reward."

Human creativity has brought forth a range of devices and furnishings to facilitate lying down. Of course, the most prominent examples are the lounge chair and the bed. Regular escapes into unconsciousness, which we usually attempt in bed, are essential to our well-being. But having the leisure to simply lie down on a lounge chair is truly a fine thing—in short, a luxury. As a piece of furniture, the lounge chair contributes to the relaxation and comfort required for contemplation—for situations in which traveling the distance to bed would feel like too much effort.

BERND BRUNNER

Common and Uncommon Ways to Lie Down

Between the short eternity that precedes the moment of birth and then follows the moment of death, beds and lounges offer opportunities for respite—for rest, sleep, and the other activities that we habitually or occasionally carry out while lying down. In fact, the time we spend horizontally adds up to at least two-thirds of our lives, although the exact figure may differ greatly from person to person. After all, a reclining position is a given for some kinds of labor or sports—mine shaft repairs or the art of the luge (although neither activity is particularly relaxing). We can also look at the horizontal profession, or *grandes horizontals*, as referred in nineteenth-century Paris to the top-tier courtesans who did business under exotic or elegant names.

No standing: A hamam *in Istanbul*

Lying down is also important in the ritualized world of the *hamam,* or Turkish bath. There, after stretching and undergoing a massage and thorough scrubbing, the visitor enjoys a lengthy rest on a warm stone slab. This relaxation period is no mere luxury, for the *hamam* treatment, which removes the upper layer of the skin and loosens the muscles, can be draining or even painful. In some Muslim countries, this type of bath continues to be a social institution for people to relax and recharge.

Finding oneself in a horizontal position with nothing solid underneath sounds like a nightmare, yet it happens all the time. Swimmers lie on or just below the waterline: the buoyancy of the water counteracts their mass, an effect that is strongest in water with a high salt content. But it remains to be seen if a horizontal shower recently touted by a Swiss manufacturer will catch on. Bathers can use this shower while lying on the stomach or on the back. *The Guardian* was not impressed, however, dismissing the device as "an absurd contraption," "sheer stupidity," and "the worst invention ever." The idea that in the future, showering could force the British

Floating like an embryo:
Dittman's patented rocking basin

BERND BRUNNER

nation to "writhe helplessly like beached seals on a platter of dead skin cells and tepid body fluids" filled the editorialist with dread.

How can we distinguish lying down from other horizontal states? Floating, for example? Roland Barthes once described floating as living "in a space without tying oneself to a place." Whether we float thanks to the absence of gravity or to the presence of intoxication, the sensation itself is always very real. And whether we are in water or in the air, floating requires neither a direction nor a destination. This very lack of purpose underlies its use as a therapeutic treatment; adherents believe that suspension in a salt-water bath kept at body temperature brings profound relaxation, even happiness.

Hard fact: Horizontal stiffening under hypnosis

Another form of floating may look like lying down to outside observers but is the result of a more mysterious process. Under hypnosis, people can take on a stiff horizontal posture. In this state, the subject appears to lie comfortably like a board across the backs of two chairs. The sight of someone in such an unusual position raises many questions: Why doesn't the body bend without a solid surface beneath it to hold it up? What does the reclining subject feel? Does he sense a force supporting him, or are his thoughts somewhere else entirely? The situation is analogous to astronauts drifting horizontally in space: while it looks as if they are reclining, the absence of any resistance to their bodies means that lying down is technically impossible. Once again, we encounter the paradox where splaying out in a horizontal position does not automatically equal lying down.

Lying Down in the Great Outdoors

Certain outdoor spots practically cry out for us to lie down on them: a lawn, a beach, or a warm rock, especially when it's been warmed by the sun. Outside, no ceiling limits our gaze, which can lose itself in the azure sky or in the movements and shifting forms of the clouds. Our view seems infinite, though depends on the positions of our bodies, whether we are lying on our backs, our sides, or our stomachs. If it's the latter, we look just over the tips of the grass and can only imagine the sky.

Sky gazing expands the soul

It's not just the lack of a boundary above us that makes lying down outside completely different from lying down in an interior space. Outside, we are subject to a wealth of sensory impressions: bright light, wind, the chirping of birds, the scent of flowers, the rushing of water, the sometimes pleasant and sometimes disturbing sounds of people and machines, approaching footsteps, a distant call suddenly breaking the silence. On the beach, we can hear how the rhythm of the waves breaks time into small, regular intervals. Even if no one else is nearby, we aren't alone. Everything, alive or inanimate, reaches us and may even speak to us if we are willing to listen. When we lie down in a landscape, it becomes as much a part of us as we become a part of it.

Lounging outdoors can have a private or public character depending on whether we are on a property shielded from others' view or a beach accessible to anyone. A legendary German beer commercial shows a man in the dunes who lets himself fall into the sand and then stretches out his arms and legs— the ultimate expression of relaxation and freedom. Those without their own yards or nearby coasts have to make do with parks. Alternatively, a field can offer a good spot to enjoy nature from a horizontal perspective, as well as protection from prying eyes.

On a hot summer night, when the air seems to

Even if you don't have your own yard,
you can still sleep outdoors

BERND BRUNNER

hang motionless in the bedroom, the temptation to flee outside with a mattress and sheet can be overwhelming. The reward would be a fresh breeze, a chance to see the stars, and perhaps a concert of birds as a wake-up call the next morning. Such an experience can be idyllic, but we should refrain from infusing the scene with an overly romantic glow. If you have ever tried to sleep outside, you know how irritating the unfamiliar sounds can be. The background noises are too different from the familiar acoustic backdrops in our bedroom. And something startling is always bound to happen outside. Once we enter shallow sleep, a relatively loud noise is sure to wake us up, even in the middle of the night. Some people claim that we hear better in the darkness, perhaps because humans harbor a faculty from an earlier phase of development, though our improved hearing in the darkness also reminds us of our vulnerability to danger when we are outside. In fact, we never seem to get used to this danger. Spending nights outdoors has been shown to produce heightened alertness in the homeless who sleep on city streets. It not only disturbs their sleep but also negatively affects their health.

Caution is especially called for in forests. There, blossoms may send out intense or even upsetting scents; creatures large and small may flutter around us or peer at us out of the darkness; and the ground

Clouds offer the most comfortable repose

may be damp and spongy. Nothing about it encourages a good night's sleep, despite the fact that our forests today are hardly the haunts of ghosts, witches, and dangerous beasts they were believed to be in the past. And although these dangerous spaces have been largely tamed, those who spend the night in a forest—or even just a wooded park—are viewed with suspicion. In the 1860s, those found spending the night in Berlin's famous Tiergarten park were automatically classified as "vagabonds" and "criminals," even if they had broken no other law.

In short, we might want to think twice before lying down or sleeping in the great outdoors.

BERND BRUNNER

Sun Worshippers

From a historical and cultural perspective, the present-day habit of lying out in the sun is quirky, to say the least. For centuries, a tan was a mark of poverty, the curse of those damned to work in the fields day after day. In the eyes of the more refined, sun-browned skin indicated a lack of cultivation. Many developments had to come together before a tan became a sign of beauty. A sun cult evolved slowly, arguing that enjoyment of the sun of went hand in hand with an appreciation of outdoor movement and exercise. The German author and mathematician Georg Christoph Lichtenberg (1742–1799), known for his aphorisms, declared that sunshine was "the primary means to promote health and vitality." He recommended sunbathing naked and once recorded his own experience of turning "nearly black" in the ocean air. Another early fan of tanning was Arnold Rikli (1823–1906). A Swiss proponent of natural medicine, he had the habit—horrifying to many of his contemporaries—of lying naked in the sun. Rikli opened the first therapeutic sunbathing facility in Slovenia in 1854. He showed that sunbathing could influence one's sense of well-being, and he increased the awareness and popularity of this practice like no one before him. History knows him as the sun doctor. In an age before anyone guessed that spending long periods in the sun could lead to skin cancer, tanning was an innocent pleasure.

The Proper Way to Lie Down

"The resting place should neither be completely horizontal nor excessively sloped," writes Isidor Poeche in his 1901 book *Sleep and the Bedroom*, which bears the unwieldy subtitle *A Hygienic-Dietetic Handbook as a Guide to Achieving Natural and Regenerating Sleep*. According to Poeche, a full horizontal position brings the risk of stroke, "especially for those with a short neck and a head that sits low, wedged between the shoulders." The reason, he argues, is that "lying completely flat facilitates the flow of blood to the brain, which is already stronger in sleep than during our waking hours." On the other hand, if the bed is too slanted, "the inconvenience easily arises that the body, which makes no intentional movements during sleep, follows the physical laws of gravity like any other lifeless mass; in other words, it tends to fall toward the center of the earth so that, instead of sleeping on the straw, one can awaken on the naked ground."

As if unaware that we cannot control our movements during sleep, Poeche warns that "neither sleeping on the back nor on the stomach is healthy, but rather damaging." He also connects sleeping in a supine position with unpleasant dreams. The up-shot is that "one should sleep on one's side, specifically the right." The ideal posture is

> somewhat curled, free of any force or pressure, in a position that allows our muscles and limbs to fully relax. A completely straight position produces as much tension as a tightly curled one, and because both variants involve effort, they prevent us from completely achieving the purpose of sleep. The feet, abdomen, and chest must be horizontal, but the head must lie approximately a half-foot higher.

Actually, none of the typical sleeping positions can be considered unhealthy; even sleeping on the stomach impedes breathing only minimally. However, it is true that slightly raising the upper body can help those who suffer from respiratory problems, such as sleep apnea, to breathe more freely. In more severe cases, sufferers must turn to technical devices that support their breathing.

Today, research has provided us with a far more detailed understanding of sleep than Isidor Poeche could boast. Just the study of the movements people make while sleeping constitutes a field of research. We now know that changes in position usually occur during phases of shallow sleep and directly influence

BERND BRUNNER

how refreshing we find a bout of slumber to be. If our motor functions are affected during sleep or we lie in the "wrong" position, we can wake up feeling exhausted. In any case, it's normal to move about while sleeping; a healthy sleeper with full motor abilities changes position up to one hundred times a night. These movements follow an individual rhythm—a unique nocturnal choreography, if you will.

No consensus exists among doctors about which is better: a certain amount of activity during sleep beyond the normal level or the greatest possible relaxation. One factor in favor of relaxation is that it leads to less tension-related pain.

If sleep researchers are correct, more than half of us sleep primarily on our sides. Older people tend to sleep in this position. However, lying on the left side can be uncomfortable for those with heart conditions. As we age, sleeping on our backs leads to an increase in snoring. If a snorer consistently wakes up in flagrante delicto to turn on his or her side, the response can eventually become unconscious.

Sleeping on the stomach requires a high level of flexibility in the neck, a flexibility that tends to diminish with age. "In a prone position, the head must be turned to the side, leading to problems for those with limited mobility," says Thomas Laser, a noted German orthopedist. "One wakes up due to neck pain and, as a result, tries to avoid lying on one's stomach." Moreover, eating too much right before bedtime presents issues for prone sleepers because, as Dr. Laser explains, "pressure on the abdominal area can cause heartburn and belching if we sleep facing downward."

Whether we are asleep or awake, a mysterious impulse can suddenly signal that it's time to change positions. What triggers this need to move? The most important factor is the pressure that a particular position exerts on certain parts of the body. Gravity and one's weight compress the body at the points where it contacts the surface below. If we're on our backs, we feel the pressure most in the shoulder blades, pelvis, and heels, whereas if we're on our sides, the pressure is more apparent in the shoulder and elbow joints, the outside of the hip, and the knee. The pressure is most evenly distributed when we sleep on the stomach or on the back, because these positions maximize the area resting on the underlying surface. If part of the body is subjected to significant pressure over a long period, blood circulates there less freely. The resulting lack of oxygen creates an unpleasant sensation, causing us to correct the imbalance by changing position. This biomechanical response is not under our rational control. Without this regular release, the compressed soft tissues would develop serious circulatory problems that could, in severe cases, lead to bedsores. Sleepers are at risk of these consequences

BERND BRUNNER

if the number of spontaneous movements they make drops to three or fewer per hour. The sleeping body also tends to seek a position in which the arm and leg joints are centered, allowing the opposing muscles to relax. It's easier to find this balance when we lie on our sides. We can get a better sense of the physiological need to change position by consciously resisting it for a few moments, so that we feel how the affected part of the body "falls asleep."

By letting the feet fall slightly outward and turning the palms of the hands up, a supine yogi brings the spine into a naturally comfortable position. This relaxation spreads throughout the entire body, and the shoulder blades drop to the ground. Although yoga practitioners call this position *savasana*—the corpse pose—they rarely give a thought to the name's macabre connotations.

In most cases, we end up sleeping in whatever position is most comfortable. The postures we assume while sleeping are not under our conscious control, but they aren't random either. If we pay attention to the body, we can sense our own patterns of movement. These factors, in addition to pressure points and spatial orientation, also play a role in whether we feel comfortable. Sometimes it feels pleasanter if the thighs touch each other, while at other times we avoid this contact and the friction it produces. Or we pull a blanket over our heads to ward off a draft or the cold but then have a need for fresh air.

For some people, the act of lying down brings a highly attuned consciousness of the dimensions of their bodies or even of their very selves. The German writer Hermann Broch vividly describes this process in his celebrated novel *The Death of Virgil*:

> He rolled on his side, his legs drawn up a little, his head resting on the pillow, the hip pressed into the mattress, the knees disposed one above the other like two beings alien to him and very far off in the distance reposed the ankles and the heels as well. How often, oh, how often in the past had he been intent on the phenomenon of lying down! Yes, it was absolutely shameful that he could not rid himself of this childish habit! He recalled distinctly the very night when he—an eight-year-old—had become conscious that there was something noteworthy in the mere act of reclining.

Position as the Key to Personality

Not only have attempts been made to draw conclusions about people's personalities from the way they sleep, but entire typologies have been drawn up. According to one popular tabloid paper, the sourpuss sleeps on his stomach with his arms slightly bent over his head, so that his fingers are spread like the toes of a frog. In this analysis, such bed frogs are burdened with problems and rarely willing to take the advice of others. Manager types, on the other hand, lie on their backs and require lots of room for their arms and legs. The obvious conclusion is that such an individual is used to being in charge and giving orders. Shy people are said to sleep on their side, with their legs drawn up, in a position that resembles an embryo in the womb; rolled into a ball, they stubbornly wait for their personalities to unfold.

Fortunate souls free of all angst sleep completely relaxed on their side, and self-confident people supposedly toss and turn less while sleeping. Such theories emerge in the gray area between serious science and pop psychology. Chris Idzikowski, from the Edinburgh Sleep Centre and Advisory Service, claims that the widespread fetal position—sleeping on the side while pulling up the legs—indicates a hard shell and a sensitive core. When meeting a stranger, such a person first seems reserved but then warms up quickly. In contrast, those who lie on their backs with their arms at their sides in the soldier position are quiet, reserved, and known for their strong principles. Yearners, on the other hand, sleep on their side and stretch out their arms in front of them. Idzikowski claims that this position indicates distrust, and that once such a person has made up his or her mind it is unlikely to change. Lying on your side with your arms against your body is the sign of the typical log sleeper, an easygoing but gullible character.

Things look quite different for free-fall sleepers, who lie on the stomach and hug their pillows. They are nervous and thin-skinned. But be warned from diagnosing your fellow man or woman on the basis of such evidence. No empirical proof of these connections exists, and serious researchers refrain from such speculations.

BERND BRUNNER

So Easy a Child Can Do It

The subject of children and lying down has long been a playground for heavy-handed theorists. The famous anatomist Andreas Vesalius, who was also the court physician for Holy Roman Emperor Charles V, voiced a most remarkable hypothesis. The otherwise highly regarded Vesalius suggested that the "typical" German head shape—flat at the back and thus relatively short—was due to the fact that German mothers placed their infants on their backs. Belgians, he further reasoned, had long heads because their mothers placed them on their sides. General wisdom cautioned against laying children in bed next to their mothers. The danger that they could be smothered was too great, and horrendous statistics show that these fears were well justified.

The designers of the Children's Pavilion at the 1873 Vienna World's Fair wanted to provide a definitive answer to the question of the ideal way for children to lie down. Realistically sculptured models illustrated a clear set of oppositions. One child lies flat on his back with his head slightly raised on a pillow, legs straight, arms stretched out against his sides. The other lies on his side with his arms under his head so that, in the words of the physician H. Plass, "the lungs cannot freely expand when breathing, the circulation is inhibited, the back is hunched and all the limbs are displaced."

While the first child oozes good health and smiles in his comfortable sleep, the other wears the pained expression of someone plagued by bad dreams. Concerned parents are said to have stood in front of the exhibit and admonished their children: "You have to lie like this one, stretched out, and not huddled together like the other one!" At the same time, Plass insisted that "sustained lying in a horizontal position" was damaging for children and drew a comparison to the practice of fattening up animals by preventing them from moving. Other experts even claimed that the high incidence of child mortality was due to children's lying excessively still. Recommended countermeasures included wicker beds with fixed legs or wooden or steel beds with movable side panels. The point was to allow children to move without falling to the floor. In the nineteenth century, it was a great sign of progress that parents

A Sami hanging bassinet

BERND BRUNNER

could be admonished to provide "a separate bed for every child!"

Other cultures took different and occasionally surprising approaches. George Catlin, a lawyer who became a painter, began traveling through the American wilderness in 1830. During his journeys he observed how a group of Native Americans had tied their children stretched out on boards, with their heads supported on bolsters so that their chins could not sink and their lips would remain closed as they slept. Elsewhere, he saw babies in basketlike devices hung on tree branches. In many parts of the world children were often laid down on soft grass, animal skins, or the bare ground, sometimes with an extra support for the head. Parents in South Africa supposedly scooped out depressions in the ashes from fires and placed the children, wrapped in animal pelts, in these hollows to protect them from the night cold.

Lying Down Together

We are rarely so alone as we are in our sleep and in our dreams. When two people lie down and sleep in the same bed, they express a profound closeness. Choosing to share a bed is a ritual of coupling that symbolizes togetherness. Nothing is as intimate as pillow talk, the conversations we have in bed while lying beside each other. A double bed with a single blanket for both partners makes it easy to sleep side by side.

In bed, a couple's struggle to find the right mix of distance and closeness takes a practical form. No matter what the outcome, their bodies speak a clear language. In her book *The Secret Language of Sleep*, Evany Thomas identifies no fewer than thirty-nine possible sleep positions, from the classic Spoon to the Tandem Cyclists to Excalibur, in which the partners are almost inextricably entwined. Other options include the Zipper, in which the couple lie back to back, with their offset lower bodies touching, and the extreme Bread and Spread, in which one partner lies directly on top of the other (who somehow manages to avoid being crushed or suffocated).

It is ill advised to make hasty conclusions

BERND BRUNNER

about the quality or psychological characteristics of a relationship based on a particular sleep position. Whether your partner turns his or her back to you when sleeping has more to do with how comfortable this position is than anything else. Yet the practice of sleeping together as a couple does raise all sorts of interesting questions. How much physical contact—stomach on back, leg on leg—can and should we tolerate? Does the level of shared proximity in bed say anything about the state of a relationship? Does sleeping in separate beds mean that deep down, a couple has given up on their relationship? Is it a signal that the end has already begun? There are no simple answers.

Some lucky people never have to worry about such thoughts keeping them awake. For them, sharing a bed becomes a habit they never question. For others, sleeping in the same bed is more problematic because of snoring (which can reach one hundred decibels, approaching the sound of an engine starting), sleeptalking, rhythmic flailing of the limbs, or restless legs syndrome. These problems are sometimes triggered by inner compulsions the sleeper cannot control. In such cases, a wider bed or separate blankets can sometimes help. Some couples also decide to stop sharing the bed as they get older so they don't disturb each other's sleep.

Gerhard Klösch, an Austrian sleep researcher, has notoriously claimed that women sleep worse with a man at their side because they feel responsible for him, and that men sleep better next to a woman for the same reason. There may be some truth to this observation; it's not uncommon for heart attack sufferers to get help more quickly because their partners noticed their distress in bed.

The question of whether it's better to sleep alone or together is answerable only on a case-by-case basis. Many people can't imagine doing without this physical closeness. But even when spending the night in the same bed leads to problems, a willingness to compromise can allow couples to still enjoy falling asleep next to each other. For example, one partner can switch to a separate bed or room during the night. Zip and link beds are also an option. These are single beds that can be joined with a zipper to create a large bed when a couple craves intimacy and later separated when they prefer to have some space. For some couples, the solution is simply to sleep in the same bed but face in opposite directions. In the early 1920s, the architect Otto Bartning designed a bedroom that sounds like the setup to a joke: the beds were separated by a wall of clear glass that allowed the happy couple to sleep side by side without suffering any resulting inconvenience.

It was long taken for granted that couples would sleep in the same bed, and it rarely occurred to anyone to question the practice. For many people, it has become an issue only in the last few decades. Now we have a range of options to use as we grapple to find the right amount of nearness and distance to each other. In some cases, sleeping apart or sleeping side by side may be what makes or breaks a relationship.

Lying Down, Sleeping, Waking Up

Regular sleep—that unconscious downtime—is a physiological necessity. People who often don't sleep well suffer mental and physical effects, and become irritable and confused. A complete lack of sleep can be deadly. One extreme example is fatal familial insomnia, an exceedingly rare and to date incurable disease recognized only in the last twenty-five years. Entire families are affected by this devastating illness, which is caused by the same genetic mutation as that which triggers Creutzfeldt-Jakob disease.

But what is sleep in the first place? The scientists and philosophers of earlier times suggested astonishing answers to this question. Aristotle, for example, claimed that eating caused fumes to form in the blood vessels that collected in the brain, causing sleepiness. Later, in an age that prided itself on its grasp of chemical principles, Alexander von Humboldt explained that sleep results from a lack of oxygen. Today we understand the processes that occur during sleep much better, but we are far from having all the answers. It is clear that sleep, viewed as an interruption of consciousness, erects a kind of barrier to perception, but one that is not completely secure

BERND BRUNNER

against external stimuli. Still, it remains a mystery in many respects. The ways we spend our days and nights, our levels of activity in the phases of the day, whether these activities are physical or mental, are intertwined and interdependent variously. Friedrich Nietzsche once wrote, "No small art is it to sleep: it is necessary for that purpose to keep awake all day."

Lying down and sleeping are more than just ways to prepare for the standing, walking, sitting, and other physical activities we engage in. Jürgen Zulley, a German sleep researcher, characterizes sleeping as "a different form of being awake." Furthermore, he claims that quality, not quantity, is what matters. Still, the state of being awake requires explanation just as much as sleep does. Why are we conscious? And while we're at it, why are we alive in the first place? "You cannot imagine life without death," wrote the Italian legal philosopher Norberto Bobbio in his wonderful book *Old Age and Other Essays*. Sleep and waking are similarly inseparable.

What happens between lying down and getting up in physiological terms? The blood pressure is highest in the arteries leading directly from the heart and falls as the blood flows through the body until in the veins before the right ventricle, it is practically nonexistent. Since all our blood vessels are laid flat when we lie down and our entire blood volume is just a few inches high as a result, hydrostatic pressure accounts for just a small share of our blood pressure overall. In other words, in a horizontal posture, the heart no longer has to pump our blood "uphill" from our legs. When we lie down, the veins in the head and neck swell noticeably, and the jugular and temporal arteries pulse more strongly. Sometimes temporary headaches and confusion can occur; these symptoms worsen if the head is positioned lower than the rest of the body. When we stand up, hydrostatic pressure comes into play as the height of the liquid column changes and the blood vessels extend over a greater height range. In the arteries supplying oxygen to the head, for example, the pressure suddenly increases, while it falls in the arteries of the legs. If we stand up very quickly, the amount of oxygen reaching the brain may drop below the needed level. In severe cases, we can end up fainting.

When we lie down, sleep is usually not far off, provided we're in the right frame of mind for it. Although we can create favorable conditions for the transition between waking and sleep, we can't plan all the details; a moment comes that we can neither control nor predict. Our eyes close; functioning slows in the muscles, including those in the neck; a feeling of heaviness floods through us. Thoughts lose their definition, and we stop concentrating on them. Our sense of space dissolves, we cede control, and consciousness slips away. Falling asleep in the presence of loud noises or other external stimuli is possible only when we are truly exhausted. We need to feel that we are safe from disturbances, unpleasant surprises, and real or imaginary dangers.

Some people, especially children, are afraid to give themselves over to the night and its slumber. For those who suffer from insomnia, lying awake can become a nightmare. Edward W. Said, the noted Palestinian American literary theorist, had the habit of going to bed late and getting up at dawn. In his autobiography, *Out of Place*, he explains that he always wanted to get sleeping over with as quickly as possible: "Sleeplessness for me is a cherished state to be desired at almost any cost; there is nothing for me as invigorating as immediately shedding the shadowy half-consciousness of a night's loss than the early morning, reacquainting myself with or resuming what I might have lost completely a few hours earlier."

Sleep can be something pleasant and welcome, and full of dreams that open up new possibilities, offer solutions, and fulfill wishes; or it can present terrible nightmares. Presumably, people whose daily rhythms were not so rigidly controlled, as is often the case today, could, despite obligations and constraints, take a more flexible attitude toward sleep and the opportune times for it. They were also not subject to the constant noise that makes it difficult for so many people today to sleep through the night.

An observer can only tell if someone is awake or asleep by listening to his or her breathing. When we sleep, our breath slows and becomes more regular. The body continues to work: peristaltic movement in the digestive tract and other essential bodily functions take place uninterruptedly. In deep sleep, even hunger and thirst cease to disturb us. But as the writer A. L. Kennedy notes, the passages between waking and sleeping are sometimes fraught with peril:

> We know what a terrible place the edge of sleep can be. It is perhaps one of the quieter reasons for making love, or rather for being each other's companions in our beds—we try to be present when the people we need most have to drop into the other little death and we like to feel them there for us when we surface badly, when we are afraid and pulling the sheet up over our faces will make no difference, will not save us.

Waking up, in particular, can bring a host of unpleasant sensations—even if we never find ourselves turned into insects overnight like Kafka's poor Gregor Samsa. It seems somewhat paradoxical, but we can wake up feeling more tired than when we went to bed, and many people start the day with terrible back pain. The English scholar Robert Burton wrote that to prevent melancholy, "waking that hurts ... by all means must be avoided." But how can we ensure that we wake up free of pain? What preparations can we take? Louis XIV's morning ritual—the *lever du roi*— is legendary: members of no fewer than six levels of the aristocracy lent a hand in easing the king though the stages of waking up and getting out of bed.

Many an unhappy soul has found the secret of getting up out of bed to be a tough nut to crack. The Scottish writer James Boswell (1740–1795) was so disturbed by a feeling of heaviness when he woke that he felt confused, testy, or "dreary as a dromedary." He longed to find a treatment that would allow him to rise from bed without experiencing severe pain. Usually he could banish the stiffness he felt only by staying in bed for a long time after waking up. He imagined a pulley especially designed to gradually lift him into a standing position, but feared that it would counter his "internal inclination" and end up causing more pain. Still, he could remember times when rising from bed had been accompanied by pleasant sensations, and did not abandon hope that something could help him: "We can heat the body, we can cool it; we can give it tension or relaxation; and surely it is possible to bring it into a state in which rising from bed will not be a pain."

From a physiological perspective, not only do parts of our musculature relax significantly during sleep, but some muscles may also shorten slightly. The result is muscular imbalances that have to be corrected when we wake up. Movements like stretching or bending the arms and torso while you sit on the side of the bed are beneficial because they help restore this balance.

Waking up also affects us psychologically. In the first moments of consciousness, the surrounding room often seems unfamiliar, and it can take a few seconds before we grasp the situation and, drawing on our memories, find our place once again. These waking moments offer an ambiguity and disorientation that may be disturbing but can also be pleasurable. And they show just how shaky the foundations of consciousness can be. Our mental map reconstitutes itself step by step, and it takes a moment before our sense of self takes shape. We have no awareness of above and below, horizontal or vertical; only the surface we're lying on seems real. Then slowly the position of the bed within the room and the surrounding furniture and windows emerge. No one has ever captured the sensation of these transitional moments as well as Marcel Proust:

> When I awoke at midnight, not knowing where
> I was, I could not be sure at first who I was; I had

only the most rudimentary sense of existence, such as may lurk and flicker in the depths of an animal's consciousness; I was more destitute of human qualities than the cave-dweller; but then the memory . . . would come like a rope let down from heaven to draw me up out of the abyss of not-being, from which I could never have escaped by myself.

Perhaps, as Proust believed, our experience of the unfamiliar is particularly intense when we fall asleep at an unfamiliar time in an unfamiliar position. When it comes to understanding the possible associations lying down can have, Proust is a central figure; thanks to his heightened sensibility, seemingly everyday behavior in bed becomes a key

The bed was his world: Marcel Proust

to the remembrance of things past: "I would lay my cheeks gently against the comfortable cheeks of my pillow, as plump and blooming as the cheeks of babyhood." To take this thought a step further: phases of sleeplessness provided Proust with better access to his past. In ways not entirely known, the alternating rhythm of short phases of sleep, dreaming, and awakening undermines the vigilance of our consciousness in a kind of involuntary memory.

For some people, an odd perceptual disturbance takes place just before they fall asleep or after they wake up. Although they are lying down, they have the impression that they are moving into a vertical position, as if they were standing up. In such an "out-of-body-experience," body and mind seem to temporarily separate, a feeling that apparently results when various sense impressions cannot immediately be brought into harmony. The same thing can happen after an epileptic seizure or certain injuries.

When we stand up, things shift back into agreement with the perspective that day-to-day life demands. Consider for a moment the words of the philosopher Hans Blumenberg in his theory of the life-world:

> Standing up to assume a vertical posture does not only multiply the quantum perceptible to us, extending its perceptibility to the point where it is not yet or no longer acute. It also creates the ability to mediate with the perceived world as the organism which has become human can compare itself to others like it. The higher or, in other words, upright

person thus also sees and hears more because he can let others see and hear for him—he can delegate these activities.

But, one is tempted to add, he loses something in the process.

Awake, Napping, Asleep

The daily cycle of light and darkness provides the underlying rhythm of our sleep, but many other factors also influence when we go to sleep and how long we stay unconscious. There is no such thing as a single natural time for us to sleep. Historically, nighttime did not simply mean peace and quiet. It was also a time of danger, when being on the lookout for enemies and wild animals was imperative. Moreover, before machines and regular working hours imposed their rhythms upon us, several periods of relaxation and sleep broke up the daily routine. Periods of wakefulness after midnight were even common. Concentrating our daily sleep into a monobloc uninterrupted by waking phases is a new habit in line with a modern society in which each activity serves a specialized purpose. Looking at how certain African or Asian societies less subject to strict time rules manage their sleep schedules provides a glimpse of what it may have been like for our society when sleeping followed an older pattern: while some sleep, others get up during the night to chat or make sure the fire does not die out. Even today it's common in Japan to see people sleeping during the day—in their offices or even the subway. Not only do the Japanese tend to sleep less at night, but falling asleep in public does not carry the same social stigma it does in the West. Of course, the fact that "normal" sleep is relative does not mean that we can simply change the way we organize our downtime.

Today we know that season, climate, and weather play a role in how long and deeply we sleep. Individual factors such as age and health also have an impact. One example of how climatic conditions can influence sleep behavior is the siesta. Common in a number of Mediterranean countries, this extended afternoon nap can last two hours or even more. The desire to sleep during the day arises when high temperatures or heavy meals put a strain on the body. While the siesta is generally considered justified in warm countries, where people tend to get up early and go to bed late, inhabitants of more northern climes have long viewed it as bad for the health. Of course, such attitudes reflected a disdain for the relaxed Mediterranean lifestyle. We now know that a short siesta—the power nap—can greatly increase performance during the second half of the workday. The tireless business minds at the British company MetroNaps recognize that a napping market exists. Their "corporate fatigue solutions" make it possible to reduce environmental influences to a minimum in order for workers to enjoy a refreshing afternoon nap.

Many workaholics have taken a dismissive attitude toward sleep. Henry Ford considered it unnecessary. Alexander the Great, Napoleon Bonaparte, Thomas Alva Edison, Winston Churchill, and—somewhat less exaltedly—Silvio Berlusconi have all been celebrated for sleeping just a few hours at night. (If or when any of these famous night owls made up this downtime during the day is unknown.)

American researchers claim to have found that those who need little sleep share a particular gene. If more recent studies are correct, sleeping longer helps us lose weight. According to this research, people who skimp on sleep actually gain weight even if they consume fewer calories. Although scientists may disagree about the reasons for this correlation, the idea that we could simply sleep away extra pounds certainly has appeal.

One of the most unusual proposals for managing how we sleep can be found in *Sleep Before Midnight*, a pamphlet published in 1953 by Theodor Stöckmann. Stöckmann, a school principal, claimed to have discovered the law of natural time. According to this law, those who go to bed early need only four and a half or five hours of sleep and thus gain three awake hours each day. The trick is to let the sun govern the rhythm of the body by going to bed when it sets and getting up no later than when it rises. It's also important to avoid "artificial suns" that "trick us of darkness and sleep."

Submitting to this natural cycle kills many birds with one stone: we balance out nervous exhaustion, regain productivity, overcome sleep problems and "aversion to active life," plug the holes in our "sieve-like memory"—in short, "cure the sufferings of body and spirit." Following the principle is so important that even the loss of social life or contact with one's family is worth the price. Stöckmann ends his tract with a prophetic rallying cry: "By consciously and willingly submitting to the cosmic dictates of the sun's orbit, we must become people of the sun, children of the light."

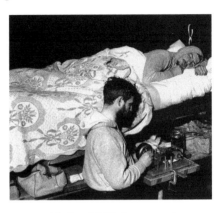

Nathaniel Kleitman and Bruce Richardson on the job

Taking the opposite approach, other researchers have asked if we can reset our sleep cycle in the absence of the sun's impact. To answer this question, the scientist Nathaniel Kleitman and his assistant Bruce Richardson retreated to a cave in Kentucky during the summer of 1938. They tried to shift their daily rhythm to a twenty-eight-hour cycle. Only Richardson was successful, but it is Kleitman who has gone down in history as a pioneer of sleep research.

Eating and Lying Down:
Better Together?

For a number of years now, lying down has enjoyed a renaissance in the form of lounging, an "activity" practiced largely, if not exclusively, in a horizontal position. Adherents gather, for example, in a softly lit room decorated entirely in orange tones, where an enormous round couch (or has it crossed the line to a bed?) invites them to get comfortable to the sound of easy-listening tunes. Some even take lounging to the next level. At B.E.D. (short for "beverage, entertainment, dining"), an elegantly designed and creatively lit restaurant in Miami Beach, patrons enjoy fusion food while lying on beds placed in different arrangements throughout the space. The menu has been planned to minimize spills: drinks are served with straws, and no soups are available. The restaurant has been open for more than ten years, though only time will tell if horizontal socializing becomes a lasting trend extending to other restaurants or even further spheres of activity. It seems likely that it will remain an exception, an expression of a particular zeitgeist targeted to specific age groups.

The ancient Greeks and Romans, on the other

hand, were known for eating while lying down. Special dining sofas would be grouped around a table. Each sofa, called a *triclinium* (from the Greek word *kline*, or bed), could accommodate as many as three men. Each would lie on his side with his head facing the table and his left elbow propped up on a pillow.

The lady of the house, other guests, or retainers of the main diners had to make do with chairs, while slaves were denied even that comfort. The Casa di Nettuno e Anfitrite in Herculaneum contains a large built-in *triclinium* that stretches from one wall to another, as does the Casa del Moralista in

Reconstruction of drinking bout in ancient Pompeii

Pompeii. The Romans lavished time on their dinner parties, starting as early as four in the afternoon.

Taking pills while lying down should be avoided because they can stay too long in the esophagus and cause damage. But there are few other practical reasons to object to horizontal dining in general. The aesthetic perspective offers a far better case against eating while lying down. The practice riles traditionalists, who see it as the end of table manners, a symptom of cultural decay, or simply as an absurdity.

Eating while sitting certainly has its advantages: the position does not limit the movements involved in eating and drinking. After all, tables and chairs have existed for thousands of years, and their anonymous inventors developed them for a reason. Sitting across from one another makes it easier to talk during the meal. And anyway, eating while reclining can easily lead to unpleasant or involuntarily comical situations: a diner's movements can result in showing a backside to someone's face or in feet inching too close to the food. Lying down may be comfortable, but it's hard to use a knife and fork in that position, and cutting up a meal into bite-size pieces can be quite a trick. It's also difficult to balance your plate on a thigh in order to cut your food or to bring your

—READING IN THE GARDEN.
SUMMER ENJOYMENT.

BERND BRUNNER

plate up to your mouth once you're back lying down. Is this really something to try without a bib? At the same time, eating in bed is possible with a backrest and a tray with a stand or folding legs.

If eating while lying down is catching on, the reason may be that many trendy restaurants feature furniture that's hard to sit on. The appropriate response to such ergonomic affronts may well be to lie down on them. Of course, there's no guarantee that lounge furniture designed first and foremost to be aesthetically pleasing will make sense ergonomically or functionally either.

Horizontal—but Hard at Work

Writers seem to have enviable lives. All they need is a pen and a piece of paper to work, or perhaps a laptop. That's what you might think, but it's not that simple, for writers are more complicated than that. To get their thoughts flowing, they need coffee, tea, cigarettes, alcohol; the right location; and the right writing position. Some writers require the background noise of a café or the lulling rhythm of a train. Others demand complete silence. Still others are notorious homebodies, content to dream about the world beyond their doors. This list is not complete.

Jean-Jacques Rousseau couldn't come up with ideas without taking a long walk. The great outdoors were his study. Just seeing a desk was enough to make him feel queasy, and working while lying down would certainly never have occurred to him. The Nobel laureate Elfriede Jelinek also needs wide-open spaces for inspiration but finds hers by looking out the window. Both writers are polar opposites of their artistic brethren who can be creative only when they lie down.

People who work while lying down often don't like to admit it. They know that their preference can quickly get them labeled as lazy. Lying down is associated with tiredness, apathy, and a lack of drive, with doing nothing, with passivity and relaxation. Goethe's industrious Faust incorporates this attitude when he declares, "If ever I lay me on a bed of sloth in peace / That instant let for me existence cease!"

Does this mean that with the exception of the occasional siesta, we should lie down only at night? Not necessarily. For some, a horizontal posture seems to create the optimal conditions for creativity and focus whatever the time of day or night.

Could it be that creativity requires a retreat from our day-to-day activities? Do artists need phases of passivity in order to make something new? There's plenty of evidence that this is true. In his letters, Marcel Proust reports that he wrote while lying down in his famous brass bed, especially during his final years, when illness confined him to his cork-lined bedroom while he completed *Remembrance of Things Past*. Everything can come to pass in bed, from erotic productivity to destructive mortality.

Proust is not the only cultural giant known for working in his bedroom. Mark Twain shared this predilection, as did Edith Sitwell, who, appropriately enough, was known for her literary portraits of English eccentrics. Lying down seems to have helped them concentrate their thoughts. William Wordsworth reportedly preferred writing his poems in bed in complete darkness, and would start over whenever he lost a sheet of paper because looking for it was too much trouble. And Walter Benjamin relates that the French symbolist poet Saint-Pol-Roux (1861–1940) wrote "LE POÈTE TRAVAILLE" ("POET AT WORK") on the door of his bedroom when he didn't want to be disturbed.

Like Proust, because of illness Heinrich Heine spent his final years in Paris writing in bed. The great German poet completed his last literary work while trapped in this "mattress grave," as he called it. W. G. Sebald, who worked on *The Rings of Saturn* while suffering back problems, lay on his stomach across his bed, propped his forehead on a chair, and placed the manuscript on the floor to write. The content of Sebald's work might be said to reflect this unenviable position. He repeatedly took up what Italo Calvino called the problem of universal gravitation, and writes of trying to achieve a state of levitation, floating on his own without external support. In *Six Memos for the Next Millennium* (1985), Calvino defines literature as "an existential function, the search for lightness as a reaction to the weight of living." If we follow Calvino a moment longer, we also learn that the state he describes can be reached only rarely because it springs from "the special connection of melancholy and humor." Could it be that he most clearly reveals the connection between lying down and creativity? Even if they are free from physical ailments, creative types don't necessarily laze around in bed for the pleasure of it. In an interview with *Le Monde*, Roland Barthes exhorted readers to "dare jj-jjjto be lazy" but confessed throwing himself on his bed with the sole purpose of "stewing" there whenever his thoughts began to circle and he felt a little down. For him at least, this phase didn't last more than fifteen or twenty minutes.

Edith Wharton, the esteemed author of *The Age of Innocence*, retreated to bed to escape rigid expectations about what women should wear. Freedom from her corset liberated her thoughts as well. She even celebrated her eightieth birthday in bed—with a candle-covered cake that caught on fire.

In an interview for *The Paris Review*, Truman Capote outed himself in a surprising manner:

I am a completely horizontal author. I can't think unless I'm lying down, either in bed or stretched on a couch and with a cigarette and coffee handy. I've got to be puffing and sipping. As the afternoon wears on, I shift from coffee to mint tea to sherry to martinis. No, I don't use a typewriter. Not in the beginning. I write my first version in longhand (pencil). Then I do a complete revision, also in longhand. Essentially I think of myself as a stylist, and stylists can become notoriously obsessed with the placing of a comma, the weight of a semicolon. Obsessions of this sort, and the time I take over them, irritate me beyond endurance.

Lin Yutang further attested to the creative benefits of lying down when he wrote, "A writer could get more ideas for his articles or his novel in this posture than by sitting doggedly before his desk morning and afternoon. For there, free from telephone calls and well-meaning visitors and the common trivialities of everyday life, he sees life through a glass or a beaded screen, as it were, and a halo of poetic fancy is cast around the world of realities and informs it with a magic beauty. There he sees life not in its rawness, but suddenly transformed into a picture more real than life itself."

Some people watch TV or listen to the radio or

music while lying down. Others read. Do some books lend themselves to horizontal reception more than others? Perhaps particular works offer a special experience—one that would be difficult to duplicate otherwise—if we read them while lying down. Do we perceive books differently in this position? Are we more susceptible to certain moods? If the theory proposed by the Argentine writer Alberto Manguel is correct, we may feel a "sense of redundancy in exploring on the page a world similar to the one surrounding us at the very moment of reading." We should balance the peaceful isolation of the couch or bed, say, with action-packed reading material. Interestingly, crime stories and horror novels are what Manguel reads to guarantee a peaceful night's sleep. Others might complain that bedtime reading like his is the best way to stay up all night. In any case, books are usually considered suitable for reading in a lounge chair when they are light and entertaining— as if too much plot would mar our vacations or weekends. Strange reasoning indeed.

Reading aside, just how much time can and should we spend in bed? According to an oft-cited statistic, we spend about a third of our lives sleeping. Another observation states that the longer we lie in bed, the longer we want to stay there. In his monumental *The Anatomy of Melancholy*, Robert Burton recommends limiting sleep to the amount that is absolutely necessary. "Nothing better than moderate sleep," he says, then adds, "Nothing worse than it, if it be in extremes, or unseasonably used." The right balance is what matters. "Waking overmuch" could be "both a symptom, and an ordinary cause," of melancholy, "yet in some cases sleep may do more harm than good, in that phlegmatic, swinish, cold, and sluggish melancholy which Melanchthon speaks of."

Nineteenth-century health gurus condemned the widespread habit of sleeping late. "The more sleep is enjoyed in moderation, the healthier it is." Groucho Marx once said that "a thing that can't be done in bed isn't worth doing at all." He was entitled to his opinion, but should we agree? Surely spending life in bed is not the answer. Muscles would atrophy, and blood circulation would slow to a crawl. Bedsores and other terrible physical consequences of excessive lying down can be seen all too clearly in people with health problems that force them to stay in a horizontal position. In 1986, eleven people spent a year in bed at the Institute of Biomedical Problems in Moscow in an effort to study the effects of a zero-gravity environment. While exercise—in some cases done next to the bed—staved off the worst consequences, the subjects required two months of physical therapy before they could sit and walk normally again.

At a certain point, spending too much time in bed becomes a problem that affects every aspect of our existence. But when do we reach this critical point? Was the famed revival preacher John Wesley right when he wrote in 1786: "By *soaking* ... so long between warm sheets, the flesh is, as it were, parboiled, and becomes soft and flabby." Wesley practiced what he preached by getting up every morning at four.

Oblomov, the main character of the 1859 novel by Ivan Aleksandrovich Goncharov that bears his name, becomes the embodiment of a person who lies around too much—to the point of doing not much else. His image has become so embedded in people's minds that it's hard to imagine the horizontal lifestyle as anything different than the one he practices. "Whenever he was at home—and almost always he was at home—he would spend his time in lying on his back. Likewise he used but the one room—which was combined to serve both as bedroom, as study, and as reception-room." Clad in a roomy oriental robe made of Persian silk, this Russian aristocrat, still in his early thirties, spends all his time daydreaming on the bed or divan. Life passes him by. "Through insufficiency of exercise, or through want of fresh air, or through a lack of both," he appears puffy and bloated, like a sausage wrapped in a dressing gown. All the attempts of his visitors to rouse him are fruitless, and even falling in love does little to change his situation. "With Oblomov," the narrator explains, "lying in bed was neither a necessity (as in the case of an invalid or of a man who stands badly in need of sleep) nor an accident (as in the case of a man who is feeling worn out) nor a gratification (as in the case of a man who is purely lazy). Rather, it represented his normal condition."

Goncharov's novel offers the defining portrait of an apathetic, even superfluous, individual who retains our sympathy thanks to his unapologetic oddness. Every age seems to produce its own Oblomov. In 1968, for example, the French filmmaker Yves Robert directed *Alexandre le bienheureux*, known in English as *Blissful Alexander*. The film tells the story of a prosperous farmer who throws his village into an uproar when his domineering wife dies and he decides to spend the rest of his life in bed. Their stories may be entertaining, but inactive characters like Oblomov and Alexandre are poor role models. Their fates demonstrate that too much time spent lying down isn't good for anyone.

Does taking drugs affect the tendency to lie down? Alcohol, at least in high doses, doesn't mix well with a horizontal posture. While they can produce a range of moods, most narcotic substances are not conducive to relaxing on the sofa or bed. Anyone who has had the highly unpleasant experience of feeling the ground sway while lying intoxicated in bed will do his best to avoid this sensation in the future. Marijuana seems a much better choice for lying down; it tends to encourage contemplation and passiveness.

Consult your doctor or pharmacist for more information on risks and side effects

Of course, we envision opium dens as full of hapless souls on the brink of unconsciousness. But smoking while lying down is a dangerous pleasure, too; it's all too easy to fall asleep with a cigarette burning. The case of Ingeborg Bachmann proves that the consequences can be fatal: the Austrian writer burned to death in 1973, when she fell asleep in her apartment in Rome while smoking. Other such stories have happier endings. Hermann von Pückler-Muskau, a German nobleman known as a vain eccentric and hedonist, used lead-based mixtures to keep his hair pitch-black well into old age. In 1828, he "saw the light" in an unforgettable way: a lamp caused his hair to catch fire. Luckily, he was able to stifle the flames by burying his head in the bedclothes. Although he lost half his hair and then decided to cut off the rest, he responded to this misfortune with humor: "Fortunately my strength does not reside in my hair." Fortunately for all of us, accidents like his are rare.

The History of the Mattress

A comfortable and restful bout of lying down requires a flexible surface on which to lie. The hips should be able to sink into whatever they're lying on; yet the waist must still be supported. When we lie on our back, the bed under us should hold up the spine without letting the pelvis sag. And to avoid hyperextending the neck, the head should not be as high as it is when we lie on our sides. If we roll over onto the stomach, the surface below must offer enough resistance to the lumbar region; otherwise, the small of the back curves in too much. Finally, if we use a pillow in this position, it should be a very flat one.

Along with a blanket and a mattress, a pillow is a necessary prop for comfortable reclining and sleeping. It serves to fill out the neck and shoulders area, hold up the head, and keep the neck flat. To avoid either pinching or overstretching the neck vertebrae, pillows shouldn't be too soft. Some people swear by neck rolls or horseshoe-shaped neck cushions. In parts of Africa and Asia, people use neck supports made of wood or stone. In some cases, sleepers who lie on their sides turn to long body pillows as a way to get comfortable.

X-rays of Egyptian mummies reveal that humans have long suffered from back problems, and that our modern lifestyle is not solely to blame for back pain. At the same time, back pain that occurs during or after a night's sleep seems to be more common now that most people don't engage in as much physical labor as they did just a few generations ago. Today, our muscles are generally not as developed or regularly used. Furthermore, people were historically smaller on average and had fewer opportunities to eat, resulting in bodily dimensions that placed less strain on the spine and in fewer people having posture problems and the illnesses such problems can cause.

Today, in contrast, the muscles that hold the torso upright take a lot of punishment, creating special demands when it comes to where we bed down for the night. The spinal disks are critical; because they are under pressure during the day, they need relief at night. A horizontal position allows them to expand. The goal of lying down is to give all the parts of the body involved in physical movement, including the muscles, a chance to relax. It's very difficult to feel well rested after a night in a worn-out bed— for example, when the spine sags between the pelvis and the shoulders because the mattress doesn't offer enough support.

The coil spring marked a genuine revolution in

sleeping surface flexibility. The approach was borrowed from other contexts: on ships and in carriages, springs softened rocking, jolting, or bumping movements and helped prevent motion sickness. The patent for coil springs was filed in 1706 in England. The first springs were wedged between thin boards, but soon it became possible to join them together to form a wire mesh. Covered with stable fabric, these constructions could be sold as elastic, "airy," easy-to-transport wire mattresses. Later, slatted bed frames added more flexibility. Switzerland's Hugo Degen and Karl Thomas are credited with this invention, which was created with solid foam mattresses in mind.

Anyone who has studied the topic of mattresses knows how complex it can be. Not all mattresses are equal, and there is a dizzying range of manufacturers and styles—not to mention the different kinds of frame and support systems. Despite the poetic associations mattresses evoke—think of all those comparisons to "floating on a cloud"—understanding the particulars of mattresses and the secrets of their design is largely a question of technical expertise.

All mattresses—whether filled with foam, latex, down, water, or the latest development, gel—adjust better to the body they support than less sophisticated sleep surfaces. Multizone mattresses that combine foam of different densities even promise to provide the right resistance for different parts of the body. Such point elasticity is particularly important when a bed is shared by two people who have different body weights or sleep in different positions and apply different amounts of stress to the same sections of a mattress. A mattress that is either too hard or too soft can cause back pain. But the sensation of softness or firmness depends on the individual. Because heavier people exert more pressure on the surface supporting them, mattresses seem softer to them than they do to lightweights. The older a person becomes, the more he or she is likely to prefer a soft mattress. Mattress retailers have no qualms about praising their products in the highest possible tones, but it takes at least a few nights to discover if sleeping on a particular mattress really feels like floating on a cloud; so trying it out in the store is practically useless. In some cases, it's hard to tell which of the salesperson's eloquent arguments are based in fact and which simply serve to shame us for our choice of bed so far. How could anyone be dumb enough to spend two decades on a no-frills feather mattress?

The box spring bed—a high bed with a mattress construction that can rest on the floor—represented a minor mattress revolution. A fabric-covered frame with steel springs, the box spring provides cushioning support for the mattress itself, which also contains a spring mesh. There's no need for slats or even

*Each after his own fashion: a fakir
with his favorite mattress*

a bed frame, and the combination of mattress and box spring gives exceptional comfort.

Those willing to make the investment can choose a particular spring and even tailor the structure of the mattress to their sleeping habits. People who sleep on their sides, for example, can select a softer design for the shoulder area and, in theory, enjoy improved circulation and a better night's sleep. But no matter how sophisticated the mattress may be, subjective factors play a large role in whether we feel comfortable on it. Identifying these factors is difficult, probably even impossible. For some, it's practically an article of faith that certain mattress components can be made only of synthetic materials—or, for others, natural ones—and this conviction influences how they feel about their mattresses. Can a mattress with steel springs really intensify electromagnetic fields in the vicinity?

It may come as a surprise that supersoft mattresses designed to shield the sleeper from any sort of disturbance, such as mattresses for preventing bedsores, can have negative consequences. The lack of stimulation can change one's perception of one's own body. Participants in a study of this phenomenon report highly unpleasant sensations: "My arms and hands disappeared, legs and pelvis were like a squished, shapeless mass and I felt as though I had melted," and "I felt like a ball, round and without contours, slowly revolving—and cold." On the other hand, "disturbances" in the form of a certain level of tolerable discomfort can help make lying down and sleeping pleasurable overall. Having a patient lie completely still, even if medically necessary, is not without problems. The body's movements are precisely what enable us to sense ourselves as a whole. Preventing such movements can result in abnormal sensations, coordination problems, and even severe identity crises.

In the universe of mattresses, water beds hold a special place. Although we often associate water beds with the 1960s, the first experiments with water-filled mattresses took place in the early nineteenth century. Neil Arnott, a Scottish physician, invented a "hydrostatic bed for invalids," a water-filled basin topped with a fabric-covered slab of rubber, as a way to provide bedridden or other highly sensitive people with a flexible, largely pressure-free surface to lie on. It's easy to predict the resulting problems. Wouldn't the weight of the patient affect the pressure in the basin and cause it to overflow? How could the water be heated so the bed wouldn't be too cold? It was many years before these critical questions could be answered.

In his 1961 cult novel, *Stranger in a Strange Land*, the science fiction writer Robert A. Heinlein described a pump system for water beds that adjusts their water levels as well as a thermostat to control the temperature. Although his design was never demonstrated in the real world, its existence was enough to prevent Charles Hall, whose 1968 prototype is considered the invention of the modern water bed, from patenting his design.

Water beds are most popular in the United States. People with problems in their joints find them comfortable. Furthermore, since they produce less dust than normal mattresses, they are a good choice for allergy sufferers. But despite a host of design improvements over the last several decades, water beds still have several disadvantages: they are so heavy that they require especially stable foundations; heating the water to body temperature uses a lot of energy; and there's always the possibility that the mattress will spring a leak.

The Archaeology of Lying Down

Like humans, animals lie down. Of course, comparisons with those that exhibit some rough physical resemblance to humans—a lounging cow chewing her cud, for example—are more useful than others. Snakes, for example, can't do anything other than lie down, even when they are moving.

Do animals need beds? Doting pet owners may provide furniture for their furry friends, but otherwise we would hardly apply the term *beds* to the places where animals sleep. At the same time, animals can go to extraordinary lengths to prepare comfortable spots to hibernate.

Unlike many animals, humans have neither fur nor especially thick skin. Since we can't expose our sensitive bodies to wind, rain, and frost, we have to make adequate preparations for sleeping. A place to sleep is a fundamental human need. But the difference between people and animals is not simply physiological. Many animals can perceive signs of danger even while sleeping and can react accordingly. In humans, these senses are not well developed. As Elias Canetti once wrote: "Anyone who lies down disarms himself so completely that it is impossible to understand how men have managed to survive sleep." Solutions to this problem can include activities such as constructing a dwelling—so in a way, human culture can be viewed as a side effect of our ancestors' efforts to get a good night's sleep.

When does a spot for sleeping become a bed? We could list criteria, such as the presence of a bed frame, a mattress, a blanket, a pillow. But the term *bed* contains a number of meanings. Although the Germanic root of the word means "a resting place dug into the earth," we use it to refer to something more comfortable. A bed is not just a provisional spot quickly organized at the side of a path, but attempts at further definition often give way to cultural prejudices. Would a typical Maasai construction of twigs covered with cowhides qualify as a bed for us? It's a tough question. But we should remember that sleeping on the floor is considered a sign of poverty in only Europe and North America; it rarely carries a stigma in other parts of the world.

Nevertheless, having some space between the ground and our bodies does have practical advantages. In addition to protecting the sleeper from dampness, which can creep into any kind of bedclothes, elevation makes it harder for insects or small animals, tame or otherwise, to pay a call. History begins at the point where memory ends. We know more about past battles and coronations than we do about the lounging habits of our forefathers. Like archaeologists studying the remnants of a forgotten language or the ruins of a temple to gain a sense of its former entirety, we can search for clues to people's lounging habits over the ages. The idea that we can fully project ourselves into their experiences is an illusion, but scattered objects and thoughts from the anonymous history of lying down help us reconstruct and begin understanding the past. For example, in 2011 a team of geoarcheologists in South Africa made a sensational find under a projecting cliff near the coastal town of Ballito: they discovered the oldest known beds. About seventy-seven thousand years ago, early Homo sapiens was already making mats of branches, sedge, leaves, and rushes to lie on, covering them with laurel leaves to keep away insects.

These mats were not yet woven or braided. All that remains today are barely recognizable traces, which could be identified only through complex analyses. Other such discoveries are "just" thirty-seven thousand years old. Until these finds came to light, remnants of beds dating back twenty to thirty thousand years that were dug up in Spain, Israel, and South Africa were considered the world's oldest.

The oldest beds discovered in more solid form are significantly younger. In the winter of 1850, the Orkney Islands of Scotland were pummeled by a powerful storm. The pounding waves unearthed parts of a Neolithic settlement, soon given the evocative name Skara Brae, that had been buried beneath a dune. Some of the five-thousand-year-old houses were preserved up to the roofs. Because wood has always been scarce in the area, the "furniture" in these ancient homes was made of stone, a boon for later historians. Some of the objects found were clearly identifiable as beds. Placed next to the hearth in the middle of the house, these generally rectangular masonry compartments of various sizes jut out into the room. Presumably they were filled with straw or heather and covered with animal skins. In other houses, the beds are built directly into the walls. We can only speculate about the inhabitants' sleeping practices. Perhaps the men slept curled up in the longer beds with their knees to their chests, or perhaps women and children shared them.

In general, early humans likely prepared places to sleep from stone, wood, or earth and covered them with layers of fur, leaves, grass, moss, or straw to make them more comfortable. As long as humans have existed, they have slept, but lying down to sleep was by no means always a given. In fact, our ancestors probably slept while squatting. In many Stone Age graves, the skeletons were found in a squatting position with drawn-up legs, sometimes bound together. It seems plausible that these early humans viewed death and sleep as related states, death being a special form of

BERND BRUNNER

sleep and vice versa. The squatting position could have been due to a lack of space in the caves and other places where these early humans sought refuge for the night and to the fact that they slept bunched together to protect themselves from wild animals and other dangers. Squatting down, they might have been more alert to the sound of an approaching threat than if they had stretched out. Viewed in this light, lying down to sleep is a sign of the progress of civilization. Sleeping outside was only possible in certain regions and at certain times of the year, so lying down to sleep presupposes the appearance indoors of a relatively large room, which would require tools to build, as protection from human and animal threats.

What had changed by the time the first great civilizations arose? While the poor still made do with plain woven mats, the beds of wealthy Egyptians, Assyrians, and Persians looked a lot like ours. Reconstructions of these beds show them to have been astonishingly delicate and elegant. They consist of a frame of palm wood that slopes slightly up at the head end and is strung with a resilient weave of palm leaves, rushes, or leather straps. Furs, tapestries, or blankets were spread on top. Instead of pillows, sleepers rested their heads by using neck supports of wood, ivory, or alabaster, often fancifully shaped or even decorated with figures. These constructions protected the users' impressive hairstyles from otherwise certain destruction. In addition, the beds featured canopies and curtains that helped keep mosquitoes at bay.

Hebrew beds were similar, but the frames were made of Lebanese cedar. In colder latitudes, the struggle for warmth played a greater role in shaping the bed's development. In northeastern China, for example, people slept on platforms that were at least thirteen feet long and could be heated through a hole on the side. This multifunctional *kang* was used as a table during the day.

The people of classical antiquity or at least the more socially elevated among them are known for spending a lot of their time lying down, whether eating, writing, or receiving guests. The *kline*, a frame of wood or bronze elevated at the head and strung with bands supporting a straw mattress, originally served wealthy Greeks exclusively as a bed. Later it was also used as a place to eat. Starting about 600 B.C., Greek men would recline together and enjoy the symposium, a ritual of dining and elevated conversation. Among the Romans, beds were highly specialized: the matrimonial bed, the lower sickbed, the catafalque for the deceased, the daybed, and the sofa-evoking dining bed with a cushion to support the arm differ clearly from one another. These pieces often boasted inlays of gold or tortoiseshell. In contrast, the masses simply slept on piles of leaves covered with sheep- or goatskins. Diogenes is said to have preferred sleeping in a wooden cask filled with straw.

Germanic peoples were unfamiliar with any kind of bed-related luxury, at least until the Romans invaded their territory. Rolled up in furs, they slept on the ground. And Roman historians reported that Celts slept in holes filled with leaves. But the years brought improvements to northern European homes: people later slept in beds that were attached to the walls and held sacks of straw. More prosperous households even had pillows and covers filled with down and feathers.

In the Middle Ages, mattresses were filled with a mixture of straw and feathers. Wealthy citizens had beds made of wood shaped by lathes. Ideas for improvements often came from cloisters, where such issues received careful consideration. This seems surprising since the monks themselves slept on simple wooden cots, perhaps with straw mattresses. St. Benedict ordered his followers to sleep on sacks of straw or leaves with felt blankest and pillows. Resulting from the body hatred and self-punishment typical of the time, the bed of a Capuchin monk, which was so hard that a sleeper left no trace on it, was held up as the ideal. Even outside monastery walls, soft beds were an object of derision. Charlemagne refused to use a mattress filled with feathers, claiming such a bed would promote effeminacy. His concerns raise the perhaps unanswerable question of just how much comfort any person needs and whether accounts of such supposed ruggedness are just myths that take on an air of truth, becoming impossible to deny or doubt.

Around A.D. 1000, the Byzantines had wooden beds with high legs and raised heads. Wealthy citizens even had mattresses filled with goose down, and tapestries and furs provided additional comfort. In the Late Middle Ages, people often outfitted beds, placed in the middle of living areas, with canopies and curtains to keep insects away. Other beds were so high that lying down in them required ladders. Henry VII's bed featured a cushioning layer of straw, which was covered with linen cloths, and a thick feather comforter with additional perfume-scented

Reclinable: a state bed

blankets and a cover made of ermine fur on top. Depictions of such magnificent beds convey the impression that people tended to sit rather than lie in them. One reason for this could be that those with high social status thought being seen lying down would damage their authority. Furthermore, lying in a flat position was associated with a very specific group, the dead.

At the royal courts of Europe, bedrooms soon acquired an important role not only in private but in public life. State bedrooms, usually found next to the lord's or lady's actual bedroom, were used for receiving visitors of equal or higher rank. Permission to sit on the bed was considered a great honor. The state bed itself, elaborately formed of fine wood, stood in the middle of the room as a symbol of social status and success. The Countess of Maine (1676–1753) reportedly directed a masked ball from her bed while she was pregnant. Approaching the bed was not generally considered good form, especially when a man was visiting a woman. Such receptions were motivated by a host of reasons, from happy occasions to more serious ones, including births, weddings, and even deaths.

The magistrate M. Simon, described at length by Jean-Jacques Rousseau in his *Confessions*, cleverly used the custom of receiving visitors while lying in bed to downplay a personal handicap. In this way he prevented the indignity that he surely would have suffered if people had met him under more conventional conditions. Simon, Rousseau explains, was not even two feet tall and spoke with two different voices: a sharp, penetrating voice—the voice of his body, which sounded "like the whistling of a key"—and a bass voice—the voice of his head. Rousseau continues:

> His legs spare, straight, and tolerably long, would have added something to his stature had they been vertical, but they stood in the direction of an open pair of compasses. His body was not only short, but thin, being in every respect of most inconceivable smallness—when naked he must have appeared like a grasshopper. His head was of the common size, to which appertained a well-formed face, a noble look, and tolerably fine eyes; in short, it appeared a borrowed head, stuck on a miserable stump.

To minimize his disability, Simon held his official audiences during the morning from his bed, hiding his body under the covers. "For when a handsome head was discovered on the pillow," Rousseau says, "no one could have imagined what belonged to it."

The bed's popularity as a place for receiving guests was short-lived, and the salon soon became the preferred scene of such activities. Bedrooms returned to their primary function, sleeping, a development reflected in the rise of the expression *chambre à coucher*.

The splendid beds and opulent temples of rest once popular in certain circles had nothing in common with the nighttime environments of most people at the time. If the rural life of past ages seems romantic to us now, it's because we don't really understand what it was like. Poor people normally slept on the floor and could count themselves lucky if they had a little straw. Or they made do with a wooden bank or a chest, perhaps next to the oven. They may not have even considered this setup "uncomfortable"; after all, they didn't have much basis for comparison and often went to bed exhausted from punishing physical labor.

In early farming households, humans and livestock shared the same living space, and the animals' body heat was a source of warmth in the winter. Finding room for the laborers—the driving force of agriculture before industrialization—was a thorny issue. In addition to payment in the form of money or crops, they received food and lodging from their masters. Customs differed from region to region, but laborers usually were given spots to sleep outside the master's quarters: in the loft under the barn roof, next to the cow or horse stalls, in the milking room, even right among the stalls themselves. Often beds weren't provided, and the worker had to make do with a hard bench or a spot on the floor. In the late nineteenth century, a certain Franz Rehbein recorded his impressions of a particularly uninviting spot to sleep at a farm near Kronprinzenkoog, a

*Sultry fantasy: the bed of maharaja Sadiq
Muhammad Khan Abbasi IV of Bahawalpur*

town in the northern German region of Schleswig-Holstein: "Hardly big enough to be able to contain a bed, and neither sun nor moon shone in. It was a niche in the kitchen, void of any comfort, dark, low, drafty, dirty; a dog's den, a coffin, a Chinese trunk; as cold in winter as an ice cellar." Sometimes such retreats were referred to as sleeping platforms, but this elevated-sounding term did little to make the experience pleasant.

Beds have existed in every conceivable form and with every kind of decoration imaginable. Some offer such excesses of bad taste that they prompt us to wonder how anyone lying on them could have slept at all. Fortunately, it's usually dark when we go to bed. One spectacular example is the ostentatious bed

built by Christofle, a Parisian manufacturer of luxury goods, for the Indian maharaja Sadiq Muhammad Khan Abbasi IV of Bahawalpur. Weighing more than a ton (including 290 kilograms [about 640 pounds] of silver), it featured a statue of a female figure at each corner. As soon as the maharaja got comfortable, music began to play and the arms of the figures began to move, stirring up a pleasant breeze at the head of the bed and shooing away flies from the foot.

To understand a construction recommended by Charles de l'Orme, Louis XIII's physician, it helps to know how difficult it was in the past to effectively heat a room and that the good doctor had a horror of dying from the effects of cold that can only be called pathological. The bed itself was set in a brick housing rather like an oven. The sleeper's head protruded through a small opening with a curtain, and the structure was insulated with layers of fur. When de l'Orme was ready to go to bed, hot bricks wrapped in linen were placed along the sides and foot of the chamber. Lawrence Wright, the peerless chronicler of the history of the bed, relates that de l'Orme was gripped by a missionary zeal to promote his design as the best

Can they hear the snoring outside? Sleeping under the spell of fresh air

possible bed. Yet it still was not enough. To satisfy his need for warmth, he supposedly also put on six pairs of socks and boots lined with cotton padding before turning in for the night. Another curiosity Wright records is an enormous funnel installed over beds to channel fresh air to sleepers who chose to keep windows closed. Essentially, it was an exhaust hood in reverse.

Another common setup for sleeping is the alcove. The modern word and its relatives, the Spanish *alcoba* and the French *alcôve*, come from the Arabic *al-qubba*, which has several meanings, including "tent," "vault," and "chamber." A bed like this could easily be mistaken for a cabinet. While the details could differ, an alcove was a windowless "bedroom" separated from the main room by a door or curtain. This opening was the only source of light and ventilation. Alcoves are remnants of a time when dwellings were not yet divided into multiple rooms with different purposes. They date from the fourteenth century and were likely inspired by the heavy, box-like beds of oak or chestnut popular during the Renaissance or perhaps by the cabins in ships. Alcoves were most common in the north and west of Europe during the eighteenth and nineteenth centuries but could also be found in North America, where they were known as cupboard beds." Alcoves were often built to accommodate two sleepers, who had to climb

in using a bench at the entrance. Those inside were protected from drafts and cold and enjoyed a measure of privacy but could still keep tabs on any household members or animals in the main room. Such a setup lacked the intimacy of a real bedroom, but few people at the time could have imagined something so exotic, let alone missed it. According to contemporary accounts, sleeping in such a cabinet was not necessarily restful. To stay warm in the winter, the sleeper had to keep the door or curtain closed, and the oxygen in the small space was quickly exhausted. The result was not only a stuffy atmosphere but ideal conditions for mice and parasites. Furthermore, the space was usually so small that those inside couldn't stretch out and had to try to get comfortable in a half-sitting position. Perhaps the discomfort was worth it; if a French legend can be believed, alcoves were built to protect sleepers from wolves that could get into houses at night.

Alcoves came under criticism at the end of the eighteenth century, when people began to recognize the health benefits of fresh air, especially as a way to combat tuberculosis. Still, shepherds in the Auvergne and Pyrenees continued to use a type of alcove, called the lit clos, a sleeping cabinet on wheels. Thanks to this mobile arrangement, they could spend the night near their flocks and scare away any wolves or bears that might turn up.

The Oriental Roots of the
Art of Lying Down

Typical European furniture in the sixteenth and seventeenth centuries offered a range of solutions for basic problems but could hardly be described as comfortable in the modern sense. Nevertheless, it apparently met the needs of the time. Outside influence was necessary to make more comfortable lounging part of our modern lifestyle and add it to our day-to-day behaviors. The impulse for this development came from the East. Enthusiasm for the "Orient" left countless marks among Europe's upper classes. Drinking coffee was one example; Louis XIV's quirk of giving himself and his mistresses "Oriental" pet names was another. The British diplomat Paul Rycaut (1629–1700) was one of the first outsiders to describe the world of the Ottoman rulers in detail. For seventeenth-century Europeans, the opulent palaces with their marble floors, velvet curtains, and divans upholstered in heavy silk were indescribably exotic. In many accounts, these Ottoman interiors became stage sets for the tellers' fantasies of unbridled eroticism. But some observers had more elevated ideas. The great German writer Johann Wolfgang von

Goethe's intensive occupation with Persian poetry and his realization that Orient and Occident were inseparable inspired *West-Eastern Divan*, one of his major works.

People were drawn to a world that seemed to be the opposite of theirs. The historian Sigfried Giedion describes the typical perception that the Western lifestyle was based in effort, as opposed to the relaxation at the root of life in the East: "In the East everyone, poor and rich alike, has time and leisure. In the West no one has." And in his "An Idyll on Idleness," Friedrich Schlegel claimed "only Italians know how to walk and only Orientals how to repose."

The Oriental influence on furniture design first became apparent in France. In the eighteenth century, the first upholstered chairs were produced. Soon bed-chair hybrids like those we still use entered the scene, making it possible to lie down without going to bed. "Couch," "chaise longue," "canapé," "divan," "recamier," "ottomane," "méridienne," and "duchesse" were labels applied to very similar pieces of furniture. But no matter what the name, they all had very little in common with their Oriental models. They were pseudo-Turkish or pseudo-Persian, because of not only how they looked but how they were used. The art historian Lydia Marinelli points to a fundamental misunderstanding between the two cultures: "While the Orient understands the cushion

The original: divan in Topkapı Palace

as an amorphous surface on which the user actively seeks a comfortable position of his own choosing, the West attempts to tailor furniture to the body in order to support its functioning." Relaxation in the East comes from lying down or sitting with crossed legs on the floor or a cushion—no armrest or backrest required. Europe's supposed Oriental furniture followed a different principle. "The languorous chaise longue encouraged an easy intimacy, not to mention lovemaking," writes the architect and writer Witold Rybczynski. "Sofas were broad not to provide for many sitters, but to allow space for the grand gesture, the leg drawn up, the arm thrown out over the back, and for the capacious clothing of that time."

A Turkish divan is a spot for sitting or reclining; it consists of a mat on the floor or a flat ledge that can run along an entire wall. In a French boudoir, on the other hand, a *divan* means an upholstered bench, often decorated with tassels and fringe, in the middle of the room. The term can even be used for a row of chairs grouped around a raised platform. In any case, divans demanded a position consisting of equal parts sitting and lying down, one enjoyed primarily by that traditionally idle class the aristocracy.

Before long this furniture developed a reputation for encouraging laziness, slackness, and "Oriental" behavior, all thoroughly at odds with the bourgeois work ethic. Sofas were also associated with drug use. All this aimless yet unbridled sprawling about was a thorn in the side of champions of propriety, who considered a military-style upright posture a prerequisite for moral integrity. Marinelli describes the sofa as a "risky location" that "leads to indecently hiked hems and unexpected touches." At the turn of the century, the German etiquette expert Konstanze von Franken was still emphatically forbidding hosts from receiving guests "lying on the chaise longue" in her perennial bestseller *Handbook of Good Form and Fine Manners*. To head off questionable situations, she recommended allowing only older ladies to sit on the sofa at all. Any man brazen enough to take a seat on the couch was summarily dismissed as being "inappropriate" and "tasteless." For von Franken, taking a more or less horizontal position was a prerogative reserved for dandies.

Field Studies of Bedrooms and Reclining Habits

Beds have to accommodate not only human biome-chanics but also the ways people in a certain time and culture lie down. In his 1924 carpentry diction-ary, Carl Wilkens writes: "As a piece of furniture that serves the purpose of complete rest—in other words, sleep—the bed must be designed to afford the human body the state of relaxation only achievable when it lies at length, and meet all requirements of health and comfort." Rarely has the function of the bed been so clearly stated. Several decades later the philosopher Otto Friedrich Bollnow described the bed as "the place from which we rise in the morn-ing and go to our daily work, and to which we return in the evening when our work is done. The course of every day (in the normal state of affairs) begins in bed and also ends in bed. And it is the same with human life: it begins in bed, and it also ends (again, assum-ing normal circumstances) in bed. So it is in the bed that the circle closes, the circles of the day as well as that of life. Here, in the deepest sense, we find rest." The bed is the primary or innermost home within the home, a place that allows and encourages a retreat to the unconscious form of our being. Yet the bed also has a flip side: it is a site of suffering and distress. The travel writer Bill Bryson has captured the paradoxical nature of beds and the rooms that house them:

> There is no space within the house where we spend more time doing less, and doing it mostly quietly and unconsciously, than here, and yet it is in the bedroom that many of life's most profound and persistent unhappinesses are played out. If you are dying or unwell, exhausted, sexually dysfunctional, tearful, wracked with anxiety, too depressed to face the world or otherwise lacking in equanimity and joy, the bedroom is the place where you are most likely to be found.

During the last centuries, sleep was turned into a private matter and forced backstage, and a sense that the intimate activities occurring in bed were shameful or embarrassing became more acute. Beds ceased to be used for representational purposes, and by the twentieth century the only publicly visible bedrooms were those for sale in furniture stores. There, as Bollnow wrote in the 1960s, they are "placed shamelessly on display." At home, the right to enter bedrooms remained limited to the immediate family.

In his book about German homes written at

about the same time, the sociologist Alphons Silbermann demonstrates that people even maintained a mental distance from their own bedrooms. Attempts to squelch awareness of the bed went so far that people referred to it as a trap or a nest. But a few years later, when barriers around private and intimate realms lowered in Western societies, it was just a matter of time until attitudes toward the bed would undergo a fundamental shift.

The bedroom was no longer a more or less hidden annex to the home, but a location one could proudly show to guests. Since that time an explosion in bed design has taken place. Beds were expected to express something about the personalities of those who slept in them, a way to create distinctions and even garner respect. Stylish lounges and beds lent their owners an avant-garde flair. Both the "secrecy" and the "inconspicuousness of this silent piece of furniture"—Bollnow's explanation as to why so few writers took up the bed as a subject or "to what a small extent the bed seems until now to have stimulated human thought"—were of the past. In an age like our own, in which boundaries seem to oscillate almost randomly between prim reticence and compulsive disclosure, embarrassing and potentially painful situations are practically guaranteed.

Just how new are the thousands of lounges and beds unveiled year after year at international furniture shows? Do they really represent innovations, or are they just endless variations of the same thing? Can the art of lying down keep up with all these advances in design? Not everything on the market follows the well-known dictum "Form follows function." Sofas that their owners can turn into "seating landscapes" in just a few simple steps or that feature laptop stands for use in a (half-)reclining position are in demand. Forty years ago wall beds were sold as the ne plus ultra in sleeping equipment. Muscling them into their horizontal position was no easy feat. But their time has passed, and they're hard to find these days. Their disappearance is not necessarily something to be sorry about.

Beds range from fussy, plush Laura Ashley models to the sleek Jailhouse Fuck, made of prison bars of dark steel and accessorized with a pair of Smith & Wesson handcuffs. The manufacturer proudly claims that it's the world's most exciting bed. We can only hope that the amatory feats of its owners live up to the promise of this backdrop. Other beds can cost as much as a car. The Rolls-Royce of beds is a model from the Swedish brand Hästens manufactured by hand from horsehair, linen, and wood over the course of many hours. The price: $99,000.

Designer lounges and beds are coveted collectibles, of course, but artworks about nothing other than beds—and not even particularly attractive ones, we could add—can fetch a good price, too. One such piece is the installation *My Bed* by the English artist Tracey Emin, a large, rumpled, messy bed with cigarette packs and butts, used condoms, and underwear as evidence of intense use. The sight immediately calls up the image of a nicotine-addicted figure lying in its midst. *A Cama Valium*, a more original work by the Portuguese artist Joana Vasconcelos, is a bed frame covered with pill packaging that criticizes the widespread obsession with tranquilizers. Surrounded by so many tablets, who would ever wake up again?

The Typical Bed

Only a tiny fraction of humankind will ever enjoy the perfect ergonomics of the beds featured in glossy design magazines. But no anthropologist has studied the average bed, the one most typical for an inhabitant of the earth today. If we compared beds in the favelas of Rio de Janeiro, Manhattan, and any randomly selected city in India, how would they overlap? A common denominator bed would certainly not be particularly luxurious and, lacking a frame and legs, would probably rest directly on the floor. Most likely it would consist of a mattress filled with straw or foam set on a mat. Imagining such a bed can remind members of more prosperous societies of how enormously spoiled they are, even if they constantly complain that their mattresses are too hard or too soft. Let us hope that the average bed provides a pleasant—ideally, a better than average—refuge for the average person who uses it.

We can't always enjoy optimal conditions for lying down, but having to compromise can have a positive effect. After all, we can only really appreciate luxuries if we learn to live without them. By occasionally experiencing lying down in simple conditions, we enable ourselves to fully soak in the pleasure a comfortable bed or a well-designed sofa offers.

An ideal bed isn't too long and narrow, but it's also not so wide that you feel lost in it. You sink into it—but not too much. The sheets are crisp and give off a fresh scent. It stands in a small room, tidy enough to prevent the urge to clean up from keeping you awake. The best bed is the one that is there when you need it.

After serving as the traditional bed in Japan for two millennia, the futon suddenly appeared in bedrooms everywhere in the 1980s. Nowadays, however, the popularity of these cotton-filled mats has waned. They were uncomfortable, despite their manufacturers' claims that they were good for your back. The trendiness that fueled their popularity eventually faded as word of their disadvantages spread. The futons Westerners can purchase today better accommodate our ways of reclining: they are thicker than their Asian counterparts and come with low frames. As a result, they are sometimes called Asian-style beds instead of futons, and they may contain latex cores and other atypical filling materials, such as

Hard but fair:
Japanese futon

coconut fiber. In Japan, futons (actually shikibutons) are placed directly on the floor mats, which are admittedly much softer than Western floors.

It's not easy to fight our way through the complex web of feng shui and the often esoteric ideas and recommendations for interior design that the West has derived from it. Unfortunately, most of us can't read the rules in the original language. Is it really possible to apply these practices developed in China more than three thousand years ago—and, at Mao Zedong's order, forbidden there now for over half a century—to the arrangement of elements in a bedroom of a modern prefabricated house? A book by Sarah Shurety respectfully dedicated to the masters of feng shui "who died in the cultural revolution" and promising "positive energy and harmony in the home"—as long as the reader pays sufficient attention to "the position of rooms within the house, which colours to use for decorating, the choice of furnishings, and many other factors"—looks understandable enough to the average reader. According to Shurety, the bedroom, which should be at the back of the house and face southwest, represents the "real you." The bed should ideally be placed "diagonally opposite the bedroom door" with the head against a wall. This placement within the room is more important than whether the bed is aligned with a particular direction of the compass. "If you have the headboard against a window it will damage your liver," she warns. Furthermore, "If you have the headboard positioned so that half of the bed is against a window and half against the wall you will not only damage your liver but you will also become more insecure and feel less supported by one of your parents." A bed with the foot facing the door is said to be in the coffin position and "drains away your energy slowly but surely (especially if the door enters into an en suite bathroom)." Shurety also recommends buying a new bed before the start of each new cycle of life, which occurs about every seven years, because beds "absorb more energy than most items of furniture." If buying a new bed every seven years is not possible, she suggests that you burn cones of incense around the bed and then place a dish filled with chalk under the bed for twenty-seven days. She notes that you are ill advised "to sleep with a new partner in a bed you previously shared with someone in a relationship that failed" because "it is more likely that your relationship will follow the pattern of the first." A "picture of a happy, smiling couple" provides just the right decorative touch. Do adherents of feng shui really follow all these rules?

Lying Down on the Road

Whether driven out of their homes, looking for better lives, or simply part of nomadic cultures, people around the world have always been on the move. And those who travel a lot necessarily rest and sleep in many different places. With sufficient money and leisure, travelers can try out not only new ways of life but also new ways of lying down. During their visit to Majorca in 1838, George Sand and Frédéric Chopin could only shake their heads at the beds offered to them:

> In Palma, one must be recommended by twenty influential persons announced in advance, and expected for several months before one can dare hope to avoid sleeping outside. All that anyone could do for us was to provide two small furnished, or rather, unfurnished rooms in a noxious quarter of the town, where travelers could count themselves lucky if they each encounter a cot with a mattress as thick and soft as a slate, a rattan chair, and as much pepper and garlic as they can eat.

Madame Sand was a careful observer, and nothing escaped her sharp glance—neither the wooden beds in the villas and country houses made only "of two sawhorses with two boards and a thin mattress on top" nor (when, as she wrote, she "attempted to pierce the secret of monastic life") "the very low alcoves decorated above with tiles like a burial chamber" in the dormitory of a Carthusian cloister.

The search for an appropriate spot to lie down away from home can become an existential challenge. If we aren't already familiar with places we are going, we have to expect that they'll be different from what we think, even if we saw a picture of the hotel room when we booked it. Perhaps the room is above a tavern that is open until the wee hours. Is the bed long and wide enough? In any case, thoughts of the many unknown people who have also lain, loved, sweated, or suffered in our temporary bed can make for an uneasy night.

Indeed, many people first realize the advantages of their native beds when they have to spend a night outside their own four walls, where they are forced to deal with unfamiliar and perhaps even unpleasant conditions. Not every strange bedroom invites the visitor to stay longer than absolutely necessary. Musty beds and sagging or short mattresses are just two potential problems. For some people, crucifixes or forest scenes with deer at the head of the bed cast a pall over their thoughts and turn trying to fall asleep into an epic struggle. Dubious bedside rugs constitute another common problem—generally shaggy, brightly colored, often matted-looking, they inspire efforts to keep your feet from actually touching them. A clear whiff of room deodorizer raises the question of just what situation necessitated such generous use of this supposedly refreshing scent. It's hard to keep your imagination in check. The writer Simon Winchester has his reasons for stating that in the age of mass tourism, the best way to travel is in your own easy chair, bed, or bathtub.

Reclining in a full airplane, at least in the economy class, is by definition a compromise. As soon as you lean your seat slightly back, you intrude into the already limited space of the person behind you. At the same time, your own physical and psychological well-being at the end of a long trip primarily depends on how comfortable you are, how far back you can tip your seat, how far you can stretch your legs—in short, whether the seat provides a pleasant environment even under difficult conditions. Frequent fliers share detailed information on-line about which airlines have the best seating and how they stack up in terms of price. Considering the aggression that a sudden reclining motion can unleash, it is amazing that there seem to be no official regulations about what passengers should be prepared to tolerate from those in seats in front of them. Do other laws prevail when we are airborne?

Prominent travelers are notorious for indulging their whims even when away from home—and getting mightily on the nerves of hotel staffs in the process. When on tour, the famous tenor Enrico Caruso supposedly insisted on always having three mattresses and no fewer than eighteen pillows, evoking comparison to the princess and the pea with her twenty mattresses and just as many eiderdown comforters. Gustave Flaubert's travel companion Maxime Du Camp ascribed to the great writer a special fondness for lying down while traveling: he "would have liked to travel, if he could, stretched out on a sofa and not stirring, watching cities, ruins, and landscapes pass before him like the screen of a panorama." Flaubert loved the idea of traveling and his memories of his adventures but was less enamored of the experience of traveling itself.

Others visited exotic places but devised strategies for taking in new scents and sounds without the risk of getting lost in crowds. In 1870, the writer Clara Mundt, better known under the pen name Luise Mühlbach, observed the changing scenes of Cairo while lying in her secure room at the famous Hotel Shepheard, which had served as Napoleon's headquarters and catered to guests with French chefs and Swiss maids. Her hotel bed served as a base within an

BERND BRUNNER

unlikely theater box, from which she could observe life playing like a film outside the windows:

> How delightful was my mood as I lay on my magnificent, comfortable bed and let the surging life of the streets flow past through the half-opened blinds. The boys driving the donkeys cried louder than their donkeys themselves, long trains of camels loaded with boards and beams panted and screeched, the ladies of the harem drove past in luxurious carriages, with the *sais* [who cleared the way] running ahead of them and the fat ugly eunuchs in their European garb on either side of the coaches.

The spirit of invention brought forth numerous modes of travel that did not require a change of bed and thus offered a measure of protection against unpleasant surprises. Early on, prosperous people devoted considerable resources to having made lounges and beds that could be easily taken along on a trip. Collapsible beds have existed since the time of Charles the Bold of Burgundy, and transportable furniture played a large role in the conquest of the American West. Lady Mary Wortley Montagu (1689–1762), the wife of a British ambassador to the Ottoman Empire, had a folding bed along for her trip from London to Constantinople. And the dowry of Queen

Ulrika Eleonora of Sweden (1688–1741) consisted in part of a silk-curtained canopy bed that could be taken apart for easy transport. Another example, this one from the second half of the nineteenth century, was the Dormouse or Mayer coach, named for the manufacturer, J. A. Mayer, of Munich. It was a roomy horse-drawn carriage with an interior that could be converted into beds. A Mayer coach was easy to identify thanks to the ventilation slits that funneled fresh air to those sleeping inside. Passengers slept while the coach was parked for the night or even, if necessary, during all-night runs. The long tradition of mobile sleeping units continues today in the Rolling Hotel. Designed by an inventive Bavarian entrepreneur, this conveyance combines a bus and a trailer with numerous sleeping cabins.

Those looking for a special thrill can even sleep in the beds of the famous—or infamous. Baghdad's Saddam Museum displays not only the former leader's weapons, uniforms, and other possessions seized when his palace was invaded in 2003 but also his bedroom. Couples can spend the night there for the equivalent of about $220. The neoclassical Villa Torlonia in the middle of Rome also draws tourists. From 1925 to 1943 Benito Mussolini called it home. Many visitors are particularly interested in il Duce's pompous bedroom, even though it's not available for overnight stays.

The increasingly popular practice of couch surfing makes it possible to find free accommodations in other cities or countries. It also fulfills a desire to experience a new place from the perspective of the people who live there instead of from a hotel. This new type of transitory, nomadic lifestyle is not, however, for the faint of heart; once in a while, the "couch" may turn out to be the host's bed.

Strange Bedfellows

In some regions of Africa and Asia, more than two or three people who are neither related nor intimate sleep together without unleashing a scandal or even requiring an explanation. These days in Botswana or the Congo, for example, it's not uncommon for people to sleep in groups. Pets may even join the mix. Communal sleeping is believed to protect against attacks by wild animals. Some cultures also believe, rather poetically, that your soul can get lost if you sleep alone.

Furthermore, the desire to be warm when sleeping is apparently so elementary that it banishes concerns about disturbances like snoring and limitations on a sleeper's free movements. In Western societies in the past, physical proximity to strangers was not considered unpleasant either; personal boundaries were drawn differently. Moreover, sharing a bed with the head of a household could even be a way to reconcile after a fierce argument.

As Danielle Régnier-Bohler makes clear in the chapter she wrote for *The History of Private Life*, nocturnal promiscuity—that is, sexual relations with varying people in bed—seems to have been par for the course in the Middle Ages and even later. In bed, that peninsula of privacy, as she calls it, people could give their feelings greater rein than almost anywhere else. It sounds a bit like a paradox: the bed was of course a private place, but everyone knew what kinds of things could happen there. Darkness invited deception or the "manipulation of reality" and was associated with guilt, adultery, and crime. Letting someone sleep alone was not only a way to grant him or her a peaceful night but a privilege and sign of honor. Another contributor to *The History of Private Life*, Philippe Contamine, explains that "sleeping together was often considered a consequence of poverty. Anyone who could afford to sleep alone wished to do so, or at any rate to sleep only with people of their own choosing."

When it comes to sleep, considerable confusion seems to have reigned in the Middle Ages: a single bed might contain couples, their children, siblings, or servants as well as soldiers, students, invalids, or the poor. Travel could bring complete strangers together in bed. This may seem odd to us, but it doesn't require extended explanation: another traveler could show up long after you had gone to bed and join you. On the other hand, if you had the misfortune to be a member of a lower class, you might have to clear out to make way for a social superior demanding a bed at your inn for the night.

*Here, there, and everywhere: sleeping room
at an inn in early modern Europe*

When many people showed up at the same place needing a bed for the night—as often happened, for example, during pilgrimages—every corner of space at an inn was precious. Sometimes the visitors were put up in the hayloft, or if straw was in short supply, indoor spots to sleep were quickly organized with hay. Terrible headaches were common after a night in such a fragrant setting. In Brussels, Albrecht Dürer once saw a bed for fifty people. It was a place for drunks to sleep off their intoxication. In his book *Dark Scenes from a Life of Wandering: Notebook of a Craftsman*, a certain D. Rocholl provides vivid stories of his nocturnal experiences, apparently in late-nineteenth-century northern Germany. He describes one inn as a "refuge for beggars, traveling

BERND BRUNNER

entertainers with and without horse and wagon, broom makers, peddlers, umbrella makers, tinkers, Slovaks, Gypsies, rascals, and all traveling homeless and idle folk, male and female alike." Soon chaos descends:

> Hardly has the straw been more or less arranged before the customers have thrown themselves onto it from all sides, many barefoot with their shoes in one hand and their stock and bundle in the other, and attempt to settle in. The shoes go under the head. Some of the most drunk want to stretch out horizontally, and only a sharp kick to the ribs persuades them to adopt a "longitudinal" posture. Skirts are quickly pulled off to cover their owners' heads. Everything happens quite quickly; the benches and tables are of course already occupied ... The alcohol fumes, the perspiration of fifty to sixty people, the smell of damp clothes, the reeking rags—what a horrendous atmosphere.

Why did separate sleeping arrangements finally catch on? Around the middle of the nineteenth century, critics began to condemn communal sleeping on hygienic or moral grounds. One frequent argument, here expressed by a French expert, was that overly close quarters brought "the bodily emissions of those involved into conflict." Such warnings also formed part of efforts to combat tuberculosis and syphilis. The origins of many illnesses remained unclear; even gout and scurvy were long considered communicable. Furthermore, inadequate ventilation could quickly lead to a lack of oxygen. When, in the nineteenth century, the construction of new living space could not keep pace with growth in the industrial centers, shift workers often had to share a bed with a bed lodger. The arrangement did not involve both parties' sleeping together, but rather taking turns using the bed.

Trying to sleep in a cold bed in an unheated room can cost even the most tired person a good night's sleep. In the past, people used hot-water bottles or sacks of warm sand to try to get comfortable. Those lacking such luxuries could make do with a brick that had been placed in the oven. Still, most people never faced the specter of freezing alone in bed. Moreover, preventing such a fate did not necessarily depend on the presence of human bedfellows. Elizabeth Charlotte, Princess Palatine, the wife of the Duke of Orleans, once wrote: "What keeps me truly warm in bed is six small puppies."

Herman Melville's *Moby-Dick* describes an instance of bed sharing that most people would happily pass up. Warned by the innkeeper that another guest will be sharing his bed, Ishmael, a sailor and the narrator, grows increasingly tense as the roommate fails to materialize. "I don't know how it is, but people like to be private when they are sleeping. And when it comes to sleeping with an unknown stranger, in a strange inn, in a strange town, and that stranger a harpooneer, then your objections indefinitely multiply." Visions of everything that could possibly happen during the night pass through his mind. As the thought of the stranger's bed linens prompts an outbreak of itching, he decides to settle down on a wooden bench but then gives up this idea, blows out the candle, and falls into bed. "Whether that mattress was stuffed with corn-cobs or broken crockery, there is no telling, but I rolled about a good deal, and could not sleep for a long time." Finally he hears heavy steps, and the stranger enters the room. "Such a face! It was of a dark, purplish, yellow color, here and there stuck over with blackish-looking squares." Just as the terrified Ishmael realizes that this fearful apparition is covered in tattoos, the "wild cannibal" puts out his light and jumps into bed. "I sang out, I could not help it now; and giving a sudden grunt of astonishment he began feeling me." At some point poor Ishmael manages to doze off, to awake the next morning—what a miracle!—with the cannibal Queequeg's arm around him "in the most loving and affectionate manner."

Ishmael remarks: "You had almost thought I had been his wife."

Every once in a while, members of Europe's and North America's highly individualized cultures, whose etiquette normally calls for avoiding physical contact with strangers, engage in communal sleeping. Under pleasant circumstances, these episodes seem less like breaches of taboo than like relics handed down from the past, awash in the romanticism of youthful camping trips and even somehow appealing. Alpine huts, with their closely packed mattresses for skiers or hikers, come to mind. Conversations fueled by beer or mulled wine can last far into the night there. In certain train compartments, sometimes at the conductor's signal, the seats are folded out to create a sleeping area that fills the entire space. Places like these bring together people who otherwise have nothing to do with one another and who continue on their different ways once the encounter is over. According to her memoirs, railroad sleeping cars were also a favorite field of operation for the French writer Catherine Millet, one of the most famous present-day pioneers of female promiscuity.

The question of whether children should sleep in a bed with adults evokes the disturbing practice of gerocomy, a curiosity of medical history that stretches back into antiquity and was publicized in England in John Floyer's *Medicina Gerocomic, or the*

Galenic Art of Preserving Old Men's Healths Explain'd
(1727). The presence of one or more youthful bodies,
the theory goes, could breathe new life into an old,
worn-out one. The idea that the child would lose a
corresponding amount of vitality does not appear to
have been a concern. Of course, the step from such
practices to sexual temptation and abuse was a small
one. It seems possible, if not likely, that gerocomical
"treatments" were just a pretext for sexual acts that
people did not discuss directly.

We'll never know whether all these welcome
or unwelcome encounters actually took place or are
just the products of active literary imaginations. But
other humans are not the only beings that can dis-
rupt our reclining and sleeping time. The most significant annoyances—bedbugs—are especially active
in summer and betray their presence with their nau-
seatingly sweet smell. Otherwise known as *Cimex lec-
tularius*, they nest in mattresses, crawl out at night,
and bite those sleeping there, leaving traces of their
saliva, which causes itching and welts. At least there's
one small comfort: they don't transmit disease.

Mechanized Reclining

The revolutions in how people lived during the nineteenth century were so fundamental and extensive that they affected nearly everything, including how we lie down. On the one hand, opportunities for workers to recline were limited to strictly defined times and places, subjecting relaxation to a rigid system of discipline. On the other hand, tireless efforts were devoted to using new technology to optimize the act of lying down and precisely track the postures of those performing it.

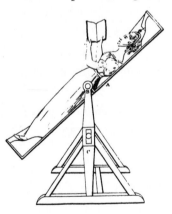

Torture or medicine? James K. Casey's "Dormant Balance"

The key question was how to relieve the back without having to lie completely flat. This search for practical hybrid forms of sitting and lying was motivated by an epidemic of back pain, as well as by the desire to help bedridden individuals whose backs suffered from constantly lying down. Doctors experimented enthusiastically with mechanical devices in the hope of curing back problems. In 1828, one James K. Casey of New York was granted a patent for a frame that could smoothly lower a patient from a vertical to a horizontal position without any effort on his or her part. Casey called his invention the Dormant Balance, and he promised that if the patient was willing to undergo this procedure two to three times per day, it could cure a crooked back. Those who wanted to spend more time horizontally were advised to use a soft mattress or other support. If we believe the illustrations that accompanied Casey's patent application, patients could relax and even read throughout this treatment.

The predecessors to flexible reclining furniture were "bed machines" with mattresses divided into separate sections for the back, thighs, and lower legs. Originally, they were joined with cumbersome wooden hinges, but by the late nineteenth century models with metal hinges had appeared. The same design principle was applied to adjustable seats in trains, hair salons, and dental practices, as well as operating tables. The designer's biggest challenge was enabling a smooth transition from sitting to lying down and back again. A surgical chair, for example, needed to accommodate all possible positions between sitting upright and lying flat. In one 1889 model, the surface consists of seven components:

a support for the head and feet, two movable armrests, and the main area itself, divided into four parts. Advertising for a metal lounge chair claimed that it was capable of seventy different positions. A mechanized reclining chair made an appearance in 1893 at the Chicago world's fair, where its manufacturer, the Marks Adjustable Folding Chair Co., enthusiastically announced, "It combines in one a handsome Parlor, Library, Smoking and Reclining chair, a perfect Lounge and a full-length Bed, and is altogether the *Best Chair in the 'Wide, Wide World.'*" Thanks to such imaginative designs, chairs not only provided padded places to sit but also made a kind of half-horizontal floating possible. A new understanding of the idiosyncrasies of sitting, lying down, and every stage in between was fueled in large part by anatomical knowledge of the hundreds of muscles involved.

A descendant of these mechanical marvels is the La-Z-Boy recliner. An old-fashioned-looking easy chair that can rotate and lean back, it has maintained its popularity for generations. Like nothing before them, these versatile lounging devices show just how much furniture belongs to us "like our skin," as the writer Hajo Eickhoff once wrote. "They form our boundary. They assume the functions of extended arms and legs, with which they complete us, and in so doing they develop typical human characteristics that we believe originate in their own being."

In addition to new technologies and materials, the nineteenth century brought a new way of seeing that triggered new forms and furniture. Movement— whether of humans, birds, or other animals—had fascinated scientists and scholars since antiquity, but never before had the laws of movement been pursued with such obsession. Eadweard J. Muybridge (1830–1904) is best known for positioning cameras at intervals and using them to photograph athletes and horses in motion. During an 1887 experiment in California, he turned his lens on clothed and nude human subjects engaged in lying down in bed or getting up again. Measured against the technical standards of the day, Muybridge's stop-motion photos represented a perceptual revolution; for the first time it was possible to see an everyday movement normally hidden from view and break it down into its separate steps.

Multifunctional furniture patented in the United States offered the middle class comfort without overloading people's typically small houses with heavy conventional pieces. A single item could

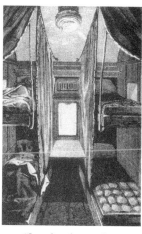

Sleeping in motion: sleeping car of the Baltimore and Ohio Railroad Company from 1847

serve as a chair, sofa, bed, and even cabinet. Beds were available that their owners could pivot horizontally and vertically, flip up, or even fold together entirely—a range of metamorphoses to offer the best possible position for any situation and the best use of limited space. One late manifestation of this trend, steel-frame sofas that turn into beds, is still common.

Does this pillow really help? Portable patented "pillow" for railroad travel

Using space optimally was an urgent issue when it came to sleeping cars on trains, and these modes of transportation proved to be a fruitful area for adjustable reclining. To develop their creations, engineers and designers drew on existing approaches in ship's cabins. Early versions left a lot to be desired in terms of comfort: occupants of the top berth were so close to the ceiling that they couldn't sit up, while those below were practically on the floor, where they could observe the feet of passersby up close. The legendary George M. Pullman obtained a patent for a smokers' sleeping car for gentlemen and a nonsmokers' counterpart for ladies. In the original version, bed compartments were hung from the ceiling, but other solutions soon appeared. Pullman's competitor Theodore T. Woodruff came up with a seating bank with a backrest that could be lowered to create two cots. Sleeping cars were exported from the United States to Europe, the original home of the railroad, in 1875. Soon "boudoir trains" were traveling between cities like Vienna and Munich. One inventive soul even developed a small portable box that travelers could easily convert into a support for the head and upper body. It doesn't seem to have caught on.

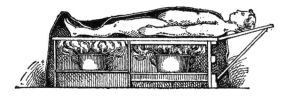

Steamed from below: recumbent sauna

Horizontal Healing

Toward the end of the nineteenth century, rest therapy became a popular way to treat those with weak nerves (today's buzzword: *neurasthenia*), hysteria, and general physical fatigue. "To provide the nervous system with the rest desirable for its recovery, we must eliminate to the greatest extent possible all sensory stimulation, efforts of the will, and arduous thought processes," wrote Leopold Löwenfeld (1847–1924), a psychologist who practiced in New York before returning to Munich, where his treatments included hypnosis. With many patients, he combined rest with the Mitchell-Playfair milk cure, a treatment named for the American Silas Weir Mitchell, who had developed it, and the English doctor William S. Playfair, who promoted it. In the most serious cases, patients had to spend six to eight weeks in bed. At the beginning of the rest period, even sitting up was prohibited, and patients were fed large amounts of milk or, later, soup or malt extract, along with a glass of champagne or red wine.

Thomas Mann soon immortalized sanatoriums and the rest cure society that took shape there by chronicling Hans Castorp's adventures in *The Magic*

Mountain. Castorp lies on an "excellent chair" on his balcony, where he can make "a proper bundle, a sort of mummy" out of himself and look forward to many satisfying hours. On the magic mountain, the horizontal condition of lying down becomes a form of being itself: "We have to lie—nothing but lie . . . Settembrini says we live horizontally—he calls us horizontallers; that's one of his rotten jokes." In his *Studies in Hysteria*, Sigmund Freud describes how after initial reluctance, he grew accustomed to "combining cathartic psychotherapy with a rest-cure which can, if need be, be extended into a complete treatment of feeding-up on Weir Mitchell lines." After all, he reasons, "This gives me the advantage of being able on the one hand to avoid the very disturbing introduction of new

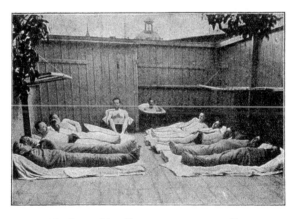

Horizontal healing: rest cure according
to Friedrich E. Bilz

physical impressions during a psychotherapy, and on the other hand to remove the boredom of a rest-cure, in which the patients not infrequently fall into the habit of harmful day-dreaming."

A form of rest cure in which patients retreat for a period into natural caves or abandoned mines continues to enjoy adherents today. The microclimate in underground shafts offers very damp, nearly dust-free air that may also be enriched with salt or radon. Spending time in such an environment can be helpful for people with asthma or other respiratory problems. And sanatoriums that provide rest cures to combat the suddenly ubiquitous problem of burnout are experiencing a boom.

To calm the nerves of sufferers, doctors once not only applied electricity directly to their bodies but sometimes set entire beds into motion. Interestingly, the vibrating or shaking bed (the second term is surely more apt) designed by Max Herz around 1900 was meant to mimic the rocking of a moving train, which had been shown to help relieve sleep disorders, general nervousness, and hearing difficulties caused by sclerosis of the middle ear. A flexible wooden board attached to a heavy base, with the help of an adjustable centrifuge device, could "be caused to vibrate like the taut string of a violin." The approach here draws on the tradition of the *fauteuil trépidant*, or vibration chair, that the noted neurologist

Jean-Martin Charcot (1825–1893) successfully used to treat patients with Parkinson's disease.

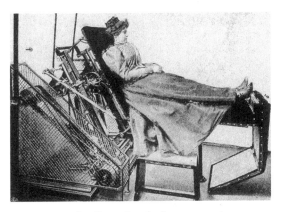

Never took off: mechanical massage therapy

Another variation was a seat designed to pound on the patient's back, intended for use in cases of chronic bronchitis (because it triggered an intense urge to cough) as well as in "rheumatic and infectious muscular processes." Hands-on treatments like these were classified under the heading "Mechanotherapy."

Floating, Rocking, Swinging

The sensation of rocking and swinging seems to make us happy. Children cannot get enough of it. And once the cradle came to be, it served as the first bed for generations of humankind. Regardless of how it was set in motion—with curved rockers, a semicircular base, or a hanging mechanism—a cradle's movement was similar to what the child experienced both in the womb and when carried by its mother and suggested closeness and comfort. A folk belief held that beechwood was most suited for making cradles because it could drive away evil spirits. Some old cradles also sported pentagrams or the letters *IHS*, the symbol of Christ, for added protection. In times of high infant mortality, such precautions surely seemed wise.

If we believe Tacitus, the ancient Germanic peoples sometimes hoisted their elderly into large

cradles hung from trees in order to rock them into the great beyond. No precise description of these remarkable beds exists, but perhaps they resembled the familiar hammock. Cradles for adults are hard to find today, but those looking to relive the sensations of childhood and perhaps even the time before birth can turn to rocking chairs and porch swings.

Caribbean hammock

During his first visit to the islands known today as the Bahamas, Christopher Columbus encountered hammocks that floated above the ground in the huts of the inhabitants. Because they were easy to fold up and transport yet offered protection from rats and snakes, they quickly became standard equipment for Spanish sailors and, later, soldiers. After that, it was only natural that they would catch on throughout the world. But a hammock is not nearly as comfortable as a reclining chair, and it limits its occupant's movements far more than sitting in a chair. Also, getting the amount of tension right when you hang it up is not always easy. Lying in a hammock can feel like being tied up, and if you roll over onto your side, you can end up falling out.

The position of a hammock in a room is not necessarily just a matter of chance and may even have a meaning. Pascal Dibie, a French cultural historian, has shown how the arrangement of hammocks among indigenous peoples in the Amazon region reflects the social relationships of those who use them. Dibie explains, for example, that in the communal rooms of the Bari "the hammocks are placed at different heights that signify the age, gender, family membership, and symbolic relationships linking their owners to one another and to the universe of the house." Young unmarried men sleep almost two meters (six and one-half feet) above the ground in the "sky" of the house and need ropes to reach their beds.

The wave of mechanization in the late nineteenth century brought changes to the hammock. For example, its netting could be reinforced with wooden slats to improve the tension. Round nets draped around the hammock could ward off pesky mosquitoes. And a blind pulled up across the entire contraption could even protect the occupant from painful sunburn.

One clever inventor hung a hammock in a large inverted tricycle and added a waterproof tentlike cover that, according to the patent, could turn a "vehicle" designed for rocking and reclining into a full-fledged bedroom. Was it a genuinely useful creation or just a dubious technical curiosity?

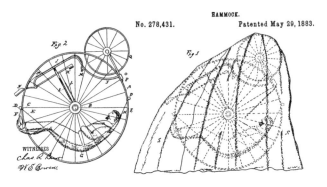

My bicycle is my castle

A hybrid of hammock and recliner—one of history's many forgotten "solutions of motion problems," in the words of Sigfried Giedion—could be found in the self-adjusting hammock chair, which was designed to be suspended from a tree. This innovation replaced the often difficult-to-handle net with a piece of canvas stretched within a frame. According to the manufacturer, this design prevents "drawing the clothing so tightly around the body, thus making it just as cool, while the annoyance of catching buttons tearing down the lady's hair, and the double somersault in the air is avoided." Although the hammock chair has not survived to the present day, it's not all that far removed from a patio swing with flowered cushions for the seat and back. If the seat is big enough, you can lie down in it as well.

Each to their own: "self-adjusting hammock chair"
for husband and wife

BERND BRUNNER

The Puzzle of the Recliner

The end of the nineteenth century saw the emergence of the reclining chair, but this new development brought its own problems. Unlike the bed, which allows its occupant to change position at any time and find a comfortable position, recliners must be carefully tailored to users' physical needs from the beginning. In 1940, Gunther Lehmann, an occupational psychologist who was highly critical of the reclining chair design of his day, formulated these requirements:

1. The area where the body lies must be as large as possible and the pressure on the surface of the parts of the body lying on it as small as possible. However, this does not mean that this pressure should be distributed evenly. On the contrary, the parts that are especially resistant to pressure (e.g., the posterior) should be subject to more of it than those that are sensitive (e.g., the lower spine). A plaster cast would not provide an ideal surface to lie upon!

2. All direct and indirect pressure on the

nerves (which causes the limbs to fall asleep) must be avoided, as must blockage of the flow back from the blood vessels in the legs, which are overly full as a result of working in a standing position.

3. The position of the limbs defined by the design of the chair must represent a state of true rest.

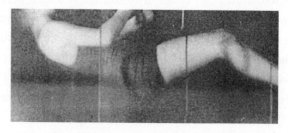

*Floating as model for recumbency:
the optimal position in the basin*

All this is more easily said than done because it takes more than cushions, arm- and footrests, and support for the head to produce that "true state of rest." Achieving it entails finding a position in which the muscles that move the hip and knee joints are completely relaxed and not subject to external forces. But how? Lehmann had the idea of submerging test subjects in a transparent basin, where they held themselves in place with a horizontal bar. He assumed that reducing the effects of gravity in this way would make it easier to identify the positions in which the muscles relax the most. Then, using photos of these experiments, he measured the angles that felt most pleasant to the participants. At 134 and 133 degrees respectively, they were nearly the same for both the hip and knee joints. Lehmann had his answer.

When we recline in a floating position, the joints are bent at these especially comfortable angles and the legs are raised. Furniture that accommodates this position existed even before Lehmann conducted his studies. They included the kangaroo sofa, inspired by its wild namesake and designed in the United States, and the *chaise longue basculante,* a recliner with a frame of steel tubing by Le Corbusier and Charlotte Perriand (1928). Seats like these are made for daydreaming.

The potted plant must stay:
"Chaiselongue Basculante," 1929

The Best Place for the Bed

When we lie down, we perceive a room's proportions, materials, and light differently from when we're standing. Our sensation of the space depends on how far the bed is from the window and whether just the head or an entire side is against the wall as well as the individual associations and memories these factors may evoke. Lying in a completely dark room without any visual points of orientation can be disturbing. Attention is then wholly concentrated on anything that stands out in the gloom: the blinking of a computer screen in standby mode; a door ever so slightly ajar; a lamp with a loose connection in the house across the street.

Many people need a little light to fall asleep easily. Some enjoy being awakened by sunlight in the morning. But not everyone enjoys falling asleep at night without the sound of traffic in the background and waking up in the morning to the melodic sound of birds. A soundproof room can make falling asleep difficult or even impossible. In such an environment, the sounds of our own breathing and digestive processes take on a whole new weight, becoming more disturbing.

In the eighteenth and nineteenth centuries, experts made it a top priority to protect people from the negative effects of almost everything. Dangers posed by lying down were no exception, and the home's surroundings, the bedroom, the materials used in beds, and even the design and position of the bed itself all came under scrutiny. The bed became a cause of powerful clashes of opinion, and there was no excuse for not knowing the consequences of sleeping incorrectly. Some recommendations still sound plausible, while others strain even the most generous amounts of credulity.

Most of this advice calls for a quiet, dark place to sleep that is protected from loud and irregular noises. However, prudence can cross the boundary into superstition, as when Isidor Poeche warns against placing the bed so that the "light falls onto the room over the sleeper's head." Such illumination, he explains, could cause children to become farsighted or cross-eyed. Sleeping with our feet toward the window is therefore preferable, although should there be no curtains or blinds, "the light stimulates our eyes and disturbs our rest." Of course, those who want to get up early can use the daylight to their advantage. The elderly are warned to make sure that the head of the bed does not face the door because the dead are carried from their rooms headfirst, a belief that echoes concerns about the coffin position in feng shui. On the other hand, the bed can intentionally be placed in this position to shorten the suffering of those dying.

Regardless of these theories, habit and the psychological aspects of our fundamental instinct to sleep in a protected place determine how people come to prefer the bed in a particular spot in the room. The anonymous author of the book *Our Household* (1964) offers a simple explanation: "Fearful individuals feel safe in niches or corners, while those with more confidence prefer the bed to stand unenclosed within the room." Janosch, famous for his children's books, offers a more modern take on this theme. He claims that his stories—"enough ideas for the next 300 years"—come to him while he is sleeping. To encourage this process, he sleeps in a variety of different places, including an attic room less than four feet high, "on a hard futon measuring six feet by six feet so that my sleeping body can turn in every direction like a compass." He explains that his body "must align itself to the stars in order to receive the transmissions. When the moon is full I sleep in a soft bed on an iron frame in a windowless room in this house, where the stone walls are three feet thick. On other days I choose a room with a small window facing east so that I don't miss the sunrise."

In *Healthy Sleep! Advice and Hints for the Unwell and the Well* (1887), Theodor Parthey ignores the effects of light to concentrate on the supposed correlation between the earth's magnetic fields and the human nervous system: "Since magnetism flows from the North Pole to the South Pole, we should ideally lie with the feet pointing south and the head toward the north so that it passes over us from the head to the foot and not in the opposite direction." The next best option, according to Parthey, is a west-east orientation "so that when we lie upright in bed, our eyes are facing south or east, but not west or north. Many a person who has fruitlessly longed for sleep has been able to remedy the situation by switching the foot and head of the bed; immediately our friend Morpheus was prepared to close his weary lids in slumber." Parthey also reports that a certain Dr. Julius von dem Fischweiler from the German city of Magdeburg—who lived for an impressive one hundred and nine years—ascribed his longevity to the fact that he always slept with his head pointing north.

Those seeking to optimize their downtime were encouraged to consider other factors. In one section of his hefty turn-of the-century bestseller *The Natural Method of Healing*, Friedrich E. Bilz describes a world overrun by enfeebled near consumptives and insists that their salvation lay in reducing the amount of dust they breathe in while sleeping. Bilz suggests climbing up on the bed frame and attaching a blanket to an open window so that it hangs over the bed. The would-be sleeper then needs to precisely align himself and the blanket to create optimal sleeping conditions. Bilz helpfully provides a diagram, which is reproduced here. According to his instructions, the trick is to lie on your back with your head propped up on a rolled-up pillow behind the blanket (at the point marked "4"). Then you grab the corners of the blanket (marked "3") and stick them under the left and right edges of your pillow (marked "5"). Thanks to these precautions, you are now completely shielded from that most dangerous of elements the air in your bedroom. Refreshing slumber is all but assured. Bilz was not the only expert making such recommendations: in an age when many people suffered from tuberculosis, prophets of well-being frequently extolled the benefits of sleeping in tents and open buildings or—in a compromise—under an open window.

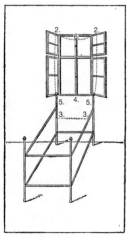

Dust-free sleeping: Bilz's diagram

Lying Down as the Stuff of Dreams—and Nightmares

During the nineteenth century, a movement gathered steam to create, in the words of the historian Peter Gay, "an age of avid self-scrutiny." The couch abetted this development. Instead of merely providing a place to lounge, it became the functional furniture of psychoanalysis, the operating table of the mind. Lying in a trancelike state on the couch, the patient grants the psychiatrist access to his or her innermost thoughts. A reclining position encourages introspection and the tendency to make playful associations and draws the inner gaze into corners and depths it does not normally reach. Desires that the patient may not consciously register during day-to-day life or may not dare express can be articulated to the psychoanalyst and then interpreted. Today, of course, the accuracy of this procedure is contested. But Freud's couch was far more than just a piece of furniture. When

Uncovering the subconscious through repose: Sigmund Freud's Ur-couch

he and his family fled from the Nazis in 1938, the couch went to London with them.

The patient stretched out on the couch and the analyst sitting upright are hardly in equal positions. Moreover, if we assume that the patient and analyst like each other or feel a charged or erotic attraction, it's easy to imagine that a sexually laden situation could result. Accounts of Sigmund Freud report that he originally sat next to the couch, where he could maintain eye contact with his patients. In response to the advances of a female patient, he moved behind the couch to head off any such situation in the future. This setup had another benefit: because patients felt less closely monitored, it was easier for them to engage in free association.

Thanks to the popularity of Freud's method, the word *couch* has taken on a whole new meaning and is now an informal synonym for psychoanalysis in many languages. When the designer Todd Bracher developed a couch for an Italian furniture company, he called his elegant creation Freud.

It seems inevitable that the psychoanalyst's couch would sometimes become the scene of intimate sharing that went beyond verbal disclosures. Freud's Hungarian student Sándor Ferenczi suffered from pangs of conscience on this account and, in a letter to his famous teacher, expressed a worry that patients with finely tuned senses might smell or even see traces of sperm on his couch. The historian Andreas Mayer has described how the couch's pornographic past caught up with its psychoanalytic present: the couch played a role in the text and even the titles of quite a few racy books. These include *Le canapé couleur de feu* (1741), supposedly by Louis Charles Fougeret de Montbron, which chronicles the sexual adventures of various church dignitaries in a Paris brothel. A number of memorable episodes play out on the establishment's sofa. Combining moral outrage and voyeuristic impulses, the text shifts "between attacking the hypocrisy of the clergy and religious education and commending its secret beating rituals as a sure cure for aging husbands," Mayer explains.

By adapting so brilliantly to the body's biomechanics, the mechanized recliners of the nineteenth century not only broadened the horizons of motion technology but channeled sexual fantasies into previously untrodden regions. It was not difficult to draw parallels between the sudden action of tipping a patient over and the abrupt movements and changes in position that occur during sex. Under the pretense of examining his female patients, the young protagonist of James Campbell Reddie's *The Amatory Experiences of a Surgeon* (1881) takes advantage of them sexually. His customized couch serves him as a "veritable battleground of Venus." He writes: "This couch was very wide, with no back, and a scroll-head at one end, whilst what would be considered the foot was half-moon shaped." Thanks to this remarkable piece of furniture, our hero is in a position "to administer [his] natural clyster . . . either standing or kneeling on a hassock." In another enthusiastic description, the narrator explains that "this couch had a most springy motion when under a pair of lively lovers, being constructed with a special eye to luxurious effect, and it also had screws at each end and at the centre, so that I could elevate the head, bottoms, or bodies of my patients to suit the ideas to be carried out." While female patients are his favorite prey, he receives a few male guests as well: "I used this couch sometimes to tie down and flagellate several of my old male patients, whose early excesses had made them too used up for the sport of love, and could only enjoy the pleasures of emission under the stimulating effects of the rod" adding that this activity "was one of the most lucrative branches" of his practice.

Suddenly, adjustable furniture was everywhere—for surgery, hairstyling, births, and physical examinations of reputable and disreputable varieties. It's therefore no surprise that these devices took up a place in the artistic imagination as well. At the turn of the nineteenth century, lying down was associated with bohemian pleasures, as well as with the perilous depths of the unconscious. As technology invaded the everyday environment and threatened to fully define it for the first time, perceptions of the horizontal world of dreams began to shift.

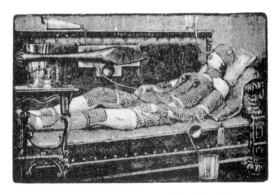

Heat treatment as art: Max Ernst,
The Preparation of Bone Glue

In their search for a visual counterpart to rationalism, the surrealists imbued sleep, dreams, and ecstatic states with aesthetic and political dimensions. Max Ernst, for example, frequently took material drawn from the world of commercial consumption and reworked it into surreal, disturbing collages. For his 1921 work *The Preparation of Bone Glue*, he added mechanical elements to the illustration of a heat treatment he had found in a medical journal. The resulting image shows a supine human figure that seems to have given up control over its surroundings. Thin tubes introduce or extract liquids from the body. It's a vision of pure horror. Does this reclining individual still possess a consciousness, or does the machine simply maintain outward signs of life? This bizarre image was intended for publication in a Dada magazine. Ernst knew that machines cannot serve as models of life and lying and that attempts to link humans and machines could turn nightmarish. His work could perhaps serve as a warning to today's technology addicts.

The Museum of Reclining

Let's take a short stroll through the Museum of Reclining, an imaginary institution housing every image ever made of a horizontal human figure. Among
the more famous works is Vittore Carpaccio's *The
Dream of St. Ursula,* which depicts its subject covered in blankets and fast asleep. Of course, not all
the individuals shown are as bloodless as Carpaccio's
holy nun; just think of Francisco de Goya's *The Nude
Maja.* Henri Matisse, who not incidentally painted
in bed using brushes attached to long sticks, shows
us a reclining nude with a most impressive backside.
He produced this painting sometime after the boom
years for such sprawling odalisques, and in the words
of the French essayist Jean-Luc Hennig, their defining anatomical feature "had tripled in size" during
the intervening period. Hennig continues: "As the
model has placed all her weight on the right side, the
buttocks appear one above the other in two storeys as
it were; it is phenomenal ... this women was heaviness that is alive. Furnished with the equipment she
has, she can probably only exist lying down, for it is
difficult to imagine her resisting the immutable laws
of gravity." Indeed, the lying figure can seem obscene,

awkward, comical but also elegant. By allowing us access to the intimate realms of their couches and beds, many of those portrayed demonstrate that lying down can be the highest form of life. And some of these horizontal hotties challenge and perhaps even arouse the observer.

Behind a curtain we find a gallery of erotic loungers. With her restrained exhibitionism, Audrey Hepburn is the epitome of stylish reclining. Her eye contact with the photographer implies more than just awareness of how attractive she is. Douglas Kirkland photographed Marilyn Monroe from above in bed, hugging a pillow and gazing lasciviously into the camera with half-closed eyes. The result—*One Night with Marilyn, Horizontal Classic*—plays with the observer's expectations, even though its desirable subject remains out of reach. In contrast, reclining men, at least those shown in bed, have always been something of

The view from horizontal: "In a frame formed by the arch of the eyebrows, nose, and mustache, a portion of my body appeared, so far as it was visible, with its surroundings."

a rarity in art. In 1972, Burt Reynolds posed on a bearskin rug for a *Cosmopolitan* centerfold. With a broad smile, a cigarette clamped in his teeth, and an arm strategically blocking the view of his groin, he is a highlight of this unusual collection. Ferdinand Hodler's *The Night* shows a very different scene: awakened by some nocturnal spirit, the artist finds himself surrounded by six other reclining figures, representing both genders and largely unclad. As we descend through the museum, we see unconscious figures displayed on the lower floors and lifeless bodies in the basement. Here, for example, we can admire Rembrandt's *Anatomy Lesson*, in which a group of doctors in training gather around a partially dissected corpse.

Are You Still Lying Down?

The French expression *être allongé,* which means "to be stretched out," is used only for people and animals, while the German verb *liegen* and the English *lie* can also apply to things. *Recline,* for its part, is something that only animate beings do. Arrente, the language of the Australian aborigines living near Alice Springs, contains the verb *ngarinyi,* which can signify not only "to lie down" but also "to sleep" and "to camp for the night" and serves as a handy euphemism for sex. Speakers of Trumai, an indigenous language of Brazil, can choose between two words for *lie*: the general *chumuchu* and the more specialized *tsula,* which refers to lying down on something other than the ground or floor. Korean also has two words to describe the act of assuming a horizontal posture—namely, *nwup-* and *cappaci-*. The first indicates a high degree of control; it is something done consciously and intentionally. The second can happen accidentally or by mistake and can also mean "to fall backwards." In Chantyal, a Tibetan-Burmese language spoken in Nepal, no fewer than seven different expressions for lying down exist. Some include specific movements, and others do not. Are we making mountains out of linguistic molehills if we ask whether lying down itself is different when it takes place in a different verbal universe? If we consider how much language shapes and influences our perceptions, then the words we have for lying down surely play a role in how we understand and perform the act.

The art of lying down is a fixture in our behavioral repertoire. It is a versatile activity that can be carried out in a variety of locations, spaces, and contexts. Anyone who has ever lived has lain down, but cycles of boom and bust are evident, too. Some epochs and cultures celebrated "cultivated," conscious reclining more than others. During the eighteenth and nineteenth centuries, Europeans looked longingly toward the East, hoping to find inspiration for their own relaxation in real or imagined practices there. While they labored under many misconceptions, their intensive search forever altered Western furniture and the ways we lie on it.

How will the way we split our time among lying down, sitting, and standing change in the future? Will we rediscover the pleasures of Roman reclining? Or unwind the way Ottomans did? Given the increasingly apparent desire for a new understanding of time, our culture seems ripe for for a new balance between vertical and horizontal existence, and ready to embrace horizontal relaxation. Lying down could play an important role in what the Korean-German cult philosopher Byung-Chul Han calls the art of lingering, part of an approach to revitalize the contemplative life and preventing existence from "declining into to a mere series of momentary presents." Some celebrities enjoy being photographed in bed, forty years after John Lennon and Yoko Ono protested the Vietnam War by staying in bed for a week. The philosopher Slavoj Žižek is among them. In fact, Žižek, dubbed an "academic rock star" by *The New York Times*, goes several steps farther. Not only does he show us his underwear and provocatively display a picture of Joseph Stalin, but he lies naked in bed and philosophizes. Žižek is the best-known protagonist of a new generation of thinkers who vaguely sense that something is off about the vertical/horizontal ratio of our lives. In his diary *Lines and Days*, the German philosopher Peter Sloterdijk challenges readers to simply "stay in bed." He writes: "You don't have to rush off into the vita activa just because the sun is already out when you wake up." It really seems as though the age of the New Horizontal has arrived, and you don't need genuine enlightenment or even cheap esotericism to explain why. This shift is a logical backlash to the compulsive idea that to get anywhere, everything must constantly be in motion. Our burned-out postmaterial society is thinking things over, and the reassessment of the horizontal is in full swing.

Of course, lying down requires no justification or complex philosophical basis. It is an actual down-to-earth activity. When we lie down, we are close to the ground and perhaps even adapt to it. The experience cultivates a bond. We let go and relax, taking a break from the constant stream of short bursts of attention that otherwise make up our days. No self-help book is ever going to teach us the "right" way to recline—we are born with an innate understanding of the grammar of horizontal orientation—but in the midst of a culture of restlessness, we can discover or rediscover the art of lying down, reduced to its bare essentials. Relaxing and stretching out all four limbs is a luxury. It is an art that we intuitively grasp, and although to understand it better, we can approach it from many angles, something irresolvable remains—a final secret, if you will. Without it, time spent horizontally would simply represent a physical state rather than a mode of living and being. Yet the art of lying down does not exist just for its own sake. It is inextricably linked to other art forms: the art of doing nothing, of being content with little, of enjoyment and relaxation and, of course, the proverbial art of love.

One of the world's largest reclining Buddhas is 150 feet long and 50 feet high. Made of plaster and brick, the statue is covered with a fine layer of gold leaf and sports intricate mother-of-pearl inlays on the soles of its feet. For more than two hundred years it has lain in the Wat Pho temple complex in Bangkok's historic district. Stretched out on his right side, the Buddha supports his head in his hand, removing pressure from his heart. He sleeps, and even in this position his vision pervades him so fully that he stays perfectly relaxed. No unrest can touch him; he is pure mystic calmness. Is he still here or already elsewhere? Let us stop for a moment at the end of our journey to take in the sight of this enviable serenity.

"What is this life if, full of care,
We have no time to lie down and stare":
150-foot reclining Buddha

For Further Reading

Barthes, Roland. *The Neutral: Lecture Course at the College de France (1977–1978)*. New York: Columbia University Press, 2005.

——. *The Grain of the Voice: Interviews 1962–1980*. Evanston, Ill.: Northwestern University Press, 2009.

Blumenberg, Hans. *Theorie der Lebenswelt*. Berlin: Suhrkamp, 2010.

Bobbio, Norberto. *Old Age and Other Essays*. Cambridge, UK: Polity, 2001.

Bollnow, Otto Friedrich. *Human Space*. London: Hyphen, 2011.

Bryson, Bill. *At Home: A Short History of Private Life*. New York: Doubleday, 2010.

Buchholz, Thomas, Anke Gebel-Schürenberg, Peter Nydahl, and Ansgar Schürenberg. "Der Körper: eine unförmige Masse. Wege zur Habituationsprophylaxe." *Die Schwester der Pfleger*, vol. 37 (July 1998), 568–72.

Burgess, Anthony. *On Going to Bed*. New York: Abbeville Press, 1982.

Calvino, Italo. *Six Memos for the Next Millennium*. Cambridge, Mass.: Harvard University Press, 1988.

Canetti, Elias. *Crowds and Power*. New York: Farrar, Straus and Giroux, 1984.

Chesterton, Gilbert Keith. *On Lying in Bed and Other Essays*. Calgary: Bayeux Arts, 2000.

Coppersmith, Fred, and J. J. Lynx. *Patent Applied for: A Century of Fantastic Inventions*. London: Coordination (Press & Publicity) Ltd., 1949.

Coughlan, Sean. *The Sleepyhead's Bedside Companion*. London: Preface, 2009.

Deutsches Hygiene-Museum (German Hygiene Museum). *Schlaf & Traum*. Exhibition catalog. Cologne: Böhlau, 2007.

Dibie, Pascal. *Ethnologie de la chambre à coucher*. Paris: Grasset, 1990.

Duby, Georges, ed. *History of Private Life, Vol. 2: Revelations of the Medieval World*. Cambridge, Mass.: Harvard University Press, 1988.

Eden, Mary, and Richard Carrington. *The Philosophy of the Bed*. New York: Putnam, 1961.

Eickhoff, Hajo. *Himmelsthron und Schaukelstuhl: Die Geschichte des Sitzens*. Munich: Hanser, 1993.

Thomas, Evany. *The Secret Language of Sleep: A Couple's Guide to the Thirty-nine Positions*. San Francisco: McSweeney's Irregulars, 2006.

Fischer, Hans W. *Das Schlemmerparadies: Ein Taschenbuch für Lebenskünstler*. Hamburg: 1949.

Freud, Sigmund. *Studies in Hysteria*. London: Hogarth Press, 1955.

Gerlach, Gudrun. *Zu Tisch bei den alten Römern: Eine Kulturgeschichte des Essens und Trinkens*. Stuttgart: Theiss, 2001.

Giedion, Sigfried. *Mechanization Takes Command: A Contribution to Anonymous History*. New York: Oxford University Press, 1948.

Goncharov, Ivan. *Oblomov*. New York, Macmillan: 1915.

Gros, Frédéric. *Marcher: une philosophie*. Paris: Carnets nord, 2009.

Han, Byung-Chul. *Müdigkeitsgesellschaft*. Berlin: Mattes & Seitz, 2010.

Hennig, Jean-Luc. *The Rear View: A Brief and Elegant History of Bottoms Through the Ages*. New York: Crown Publishers, 1995.

Henning, Nina, and Heinrich Mehl. *Bettgeschichte(n): Zur Kulturgeschichte des Bettes und des Schlafens*. Museum of Schleswig-Holstein, 1997.

Hill, Pati. "Truman Capote, the Art of Fiction No 17." *The Paris Review*, vol. 16 (Spring–Summer 1957).

Hodgkinson, Tom. *How to Be Idle: A Loafer's Manifesto*. New York: Harper Perennial, 2007.

Lehmann, Gunther. "Zur Physiologie des Liegens." *Arbeitsphysiologie* (1940), 253–58.

Levine, Robert. *A Geography of Time: The Temporal Misadventures of a Social Psychologist, or How Every Culture Keeps Time Just a Little Bit Differently*. New York: Basic Books, 1997.

Linke, Hans Robert. *Die Geschichte des Bettes und der Matratze aus orthopädischer Sicht*. Düsseldorf: self-published dissertation, 1979.

Manguel, Alberto. *A History of Reading*. New York: Viking, 1996.

Marinelli, Lydia, ed. *Die Couch. Vom Denken im Liegen*. Munich: Sigmund Freud Museum.

Munich: Prestel, 2006.

Newman, John. *The Linguistics of Sitting, Standing, and Lying*. Amsterdam and Philadelphia: John Benjamins, 2002.

Panati, Charles. *Extraordinary Origins of Everyday Things*. New York: Perennial Library, 1987.

Paquot, Thierry. *The Art of the Siesta*. New York: Marion Boyars, 2003.

Perrig, Severin: *Am Schreibtisch großer Dichter und Denkerinnen: Eine Geschichte literarischer Arbeitsorte*. Zurich: Rüffer & Rub, 2011.

Ploss, Hermann H. *Das kleine Kind: Vom Tragbett bis zum ersten Schritt*. Berlin: A. G. Auerbach, 1881.

Poeche, Isidor. *Der Schlaf und das Schlafzimmer: Ein hygienisch-diätisches Handbuch als Wegweiser zur Erlangung eines natürlichen und erquickenden Schlafes*. Berlin: E. Beyer, 1901.

Proust, Marcel. *Remembrance of Things Past, Vol. 1: Swann's Way*. Adelaide: University of Adelaide (e-book), 2012.

Rousseau, Jean-Jacques. *Confessions*. London: Aldus Society, 1903.

Rybczynski, Witold. *Home: A Short History of an Idea*. New York: Viking, 1986.

Sand, George. *A Winter in Majorca*. Chicago: Academy Press, 1978.

Selle, Gert. *Die eigenen vier Wände. Zur verborgenen Geschichte des Wohnens*. Frankfurt am Main: Campus-Verlag, 1999.

Silbermann, Alfons. *Vom Wohnen der Deutschen: Eine soziologische Studie über das Wohnerlebnis*. Frankfurt am Main: Fischer Bücherei, 1966.

Tergit, Gabriele. *Das Büchlein vom Bett*. Berlin: Herbig, 1954.

Wright, Lawrence. *Warm and Snug: The History of the Bed*. London: Routledge, 1962.

FOR FURTHER READING

Illustration and Photo Credits

ii, 24, 123, 133, 137 from Giedion, Sigfried: *Die Herrschaft der Mechanisierung: Ein Beitrag zur anonymen Geschichte* (Mechanization Takes Command). Hamburg, 2000 / 11, 58 from Wright, Lawrence: *Warm and Snug: The History of the Bed.* London, 1962 / 17 Murat Oğurlu / 19, 63, 127, 142 from Bilz, Friedrich E.: *Das neue Naturheilverfahren. Lehr- und Nachschlagebuch der naturgemäßen Heilweise und Gesundheitspflege* (The Natural Method of Healing). Leipzig, 1898 / 21 from Kuhbier, Max: *Menschen am Wasser.* Berlin, 1936 / 49 with the kind permission of culture images/Lebrecht / 55 courtesy of Mammoth Cave National Park, Kentucky, USA/Park Museum Collection / 89, 124 from Coppersmith, Fred, and J. J. Lynx: *Patent Applied for: A Century of Fantastic Inventions.* London, 1949 / 137 with the kind permission of the Fondation Le Corbusier / 143 with the kind permission of the Sigmund Freud Privatstiftung Vienna / 147 with the kind permission of VG Bild/Kunst / 150 from Mach, Ernst: *Die Analyse der Empfindungen und das Verhältnis des Physischen zum Psychischen.* Jena, 1911.

All other illustrations and photos are from the author's personal archive or a source that could not be determined.

Index of Names

Alexander the Great, 54

Aristotle, 42

Arnott, Neil, 75–76

Bachmann, Ingeborg, 69

Barthes, Roland, 19, 63

Bartning, Otto, 41

Benjamin, Walter, 62

Berlusconi, Silvio, 54

Bilz, Friedrich E., *127*, 141–42, *142*

Blumenberg, Hans, 50–51

Bobbio, Norberto, 43

Bollnow, Otto Friedrich, 96, 97, 98

Bonaparte, Napoleon, 54, 108

Boswell, James, 47

Bracher, Todd, 144

Broch, Hermann, 32

Bryson, Bill, 97

Burton, Robert, 47, 65–66

Calvino, Italo, 62–63

Canetti, Elias, 10, 77–78

Capote, Truman, 63–64

Carpaccio, Vittore, 149

Caruso, Enrico, 108

Casey, James K., *120*, 121

Catlin, George, 37

Charcot, Jean-Martin, 129

Charlemagne, 83–84

Charles the Bold, 109

Charles V, 35

Chesterton, Gilbert Keith, 12–13

Chopin, Frédéric, 105

Churchill, Winston, 54

Columbus, Christopher, 131

Contamine, Philippe, 113

Degen, Hugo, 72

de l'Orme, Charles, 89–90

Dibie, Pascal, 132

Du Camp, Maxime, 108

Dürer, Albrecht, 114

Edison, Thomas Alva, 54

Eickhoff, Hajo, 122

Elizabeth Charlotte (d'Orléans), Princess Palatine, 116

Emin, Tracey, 100

Ernst, Max, *147*, 147–48

Ferenczi, Sándor, 144
Fischer, Hans W., 10
Flaubert, Gustave, 108
Floyer, John, 118–19
Ford, Henry, 54
Fougeret de Montbron,
    Louis-Charles, 145
Freud, Sigmund, 127–28,
    143–44, *143*, 144

Gay, Peter, 143
Giedion, Sigfried, 93, 133
Goethe, Johann Wolfgang
    von, 61, 92–93
Goncourt, Edmond de, 9
Goncourt, Jules de, 9
Goncharov, Ivan
    Aleksandrovich, 67, 68
Goya, Francisco de, 149

Hall, Charles, 76
Han, Byung-Chul, 154
Heine, Heinrich, 62
Heinlein, Robert A., 76
Hennig, Jean-Luc, 149
Henry VII, 84
Hepburn, Audrey, 150
Herz, Max, 128
Hodler, Ferdinand, 151
Humboldt, Alexander von, 42

Idzikowski, Chris, 33–34

Janosch, 140
Jelinek, Elfriede, 60

Kafka, Franz, 46
Kennedy, A. L., 46
Kirkland, Douglas, 150
Kleitman, Nathaniel, 55, *55*
Klösch, Gerhard, 40

Laser, Thomas, 29–30
Le Corbusier, 9, 137
Lehmann, Gunther, 135–37
Lennon, John, 154
Levine, Robert, 5
Lichtenberg, Georg
    Christoph, 25
Lin Yutang, 5, 64
Louis XIII, 89
Louis XIV, 47, 92
Löwenfeld, Leopold, 126

Maine, Countess of, 85
Manguel, Alberto, 65
Mann, Thomas, 126–27
Mao Zedong, 103
Marinelli, Lydia, 93–94, 95
Marx, Groucho, 66
Matisse, Henri, 149
Mayer, Andreas, 145
Mayer, J. A., 110
Melville, Herman, 116
Michelangelo, 12–13

Millet, Catherine, 118

Mitchell, Silas Weir, 126, 127

Monroe, Marilyn, 150

Montagu, Lady Mary
   Wortley, 109

Mühlbach, Luise, *see* Mundt,
   Clara

Mundt, Clara, 108–109

Mussolini, Benito, 110

Muybridge, Eadweard J., 123

Nietzsche, Friedrich, 43

Ono, Yoko, 154

Parthey, Theodor, 140–41

Perriand, Charlotte, 137

Plass, H., 36

Playfair, William S., 126

Poeche, Isidor, 27–28, 139

Proust, Marcel, 48–50, 61, 62

Pückler-Muskau, Hermann,
   Fürst von, 69

Pullman, George M., 124–25

Reddie, James Campbell,
   145–46

Régnier-Bohler, Danielle,
   112–13

Rehbein, Franz, 87–88

Rembrandt, 151

Reynolds, Burt, 151

Richardson, Bruce, 55, *55*

Rikli, Arnold, 25–26

Robert, Yves, 68

Rocholl, D., 114–15

Rousseau, Jean-Jacques, 60,
   85–86

Rybczynski, Witold, 94

Rycaut, Paul, 92

Sadiq Muhammad Khan
   Abbasi IV, *88*, 89

Said, Edward W., 45

Saint-Pol-Roux, 62

Sand, George, 105–106

Schlegel, Friedrich, 93

Sebald, W. G., 62

Shurety, Sarah, 103–104

Silbermann, Alphons, 98

Simon, M., 85–86

Sitwell, Edith, 61

Sloterdijk, Peter, 154

Stalin, Joseph, 154

St. Benedict, 83

Stöckmann, Theodor,
   54–55

Thomas, Evany, 38

Thomas, Karl, 72

Twain, Mark, 61

Ulrika Eleonora, Queen,
   109–10

Vasconcelos, Joana, 100
Vesalius, Andreas, 35
von dem Fischweiler,
    Julius, 141
von Franken, Konstanze, 95

Wesley, John, 67
Wharton, Edith, 63

Wilkens, Carl, 96
Winchester, Simon, 107
Woodruff, Theodore T., 125
Wordsworth, William, 61–62
Wright, Lawrence, 89–90

Žižek, Slavoj, 154
Zulley, Jürgen, 43

Acknowledgments

My sincere thanks to Wolfgang Hörner, Peter Brandes, José João Carvalho, Olaf Dufey, Detlef Feussner, Herbert Gebhard, Petra Heizmann, Thomas Kluge (Ravi Inder Singh), Thomas Laser, Ulrich Meyer, Stanislaus von Moos, John Newman, Severin Perrig, Sebastian Posth, Florian Ringwald, Philipp Sarasin, Evelin Schultheiss, Hari Steinbach, the Sigmund Freud Museum, and my parents, Helgard and Siegfried Brunner, who first taught me the art of lying down.

My thanks also to Valerie Merians, Dennis Johnson, and Kelly Burdick of Melville House, who took an early interest in bringing my book to English-speaking readers.

Last but not least, I was delighted to work once again with my friend and frequent translator Lori Lantz, who writes exactly as I would if my native language were English.

Acknowledgments